THE SPACEFLIGHT NARRATIVE

THE SPACEFLIGHT NARRATIVE

A LITERARY HISTORY OF LEAVING EARTH

ALEXANDER MACDONALD

THE MIT PRESS   CAMBRIDGE, MASSACHUSETTS   LONDON, ENGLAND

The MIT Press
Massachusetts Institute of Technology
77 Massachusetts Avenue
Cambridge, MA 02139
mitpress.mit.edu

The MIT Press would like to thank the anonymous peer reviewers who provided comments on drafts of this book. The generous work of academic experts is essential for establishing the authority and quality of our publications. We acknowledge with gratitude the contributions of these otherwise uncredited readers.

This book was set in Stone Serif and Stone Sans by Westchester Publishing Services.
Printed and bound in the United States of America.

Library of Congress Cataloging-in-Publication Data

Names: MacDonald, Alexander C. author
Title: The spaceflight narrative : a literary history of leaving earth /
Alexander MacDonald.
Description: Cambridge, Massachusetts : The MIT Press, 2026. |
Includes bibliographical references and index.
Identifiers: LCCN 2026002885 (print) | LCCN 2026002886 (ebook) |
ISBN 9780262056830 paperback | ISBN 9780262056847 pdf |
ISBN 9780262056854 epub
Subjects: LCSH: Science fiction—History and criticism | Space flight in literature |
LCGFT: Literary criticism
Classification: LCC PN3433.6 .M33 2026 (print) | LCC PN3433.6 (ebook)
LC record available at https://lccn.loc.gov/2026002885
LC ebook record available at https://lccn.loc.gov/2026002886

10 9 8 7 6 5 4 3 2 1

EU Authorised Representative: Easy Access System Europe, Mustamäe tee 50, 10621
Tallinn, Estonia | Email: gpsr.requests@easproject.com

CONTENTS

INTRODUCTION

Science-fiction stories on interplanetary trips carry new ideas to the masses. All who work on these lines work well, arousing interest, promoting the working of the brain, and training up people who are attracted by the idea of grand projects, and who will work in this field in the future.[1]

—Konstantin Tsiolkovsky, "Is This Mere Fantasy?," 1935

We have the vehicles today to travel into space because, for hundreds of years, the evolving narratives of spaceflight have motivated people to work to make these vehicles a reality. Although the spacecraft is often considered a product of the twentieth century, it was in the period from the mid-seventeenth century to the end of the nineteenth that it first took shape in the crucible of the human mind. Long before the spacecraft existed as an object, it existed as an idea. In works of both fiction and nonfiction, the idea of the spacecraft was played with, considered, and shaped in dozens of literary works over centuries before the metal began to be poured for its physical realization. It was in literature, and through stories, that early spacecraft concepts were developed and defined. It was through books that the personal motivations for spaceflight—the stunning cosmic vistas, the adventures of exploration and technical achievement, and the potential encounters with alien cultures—were communicated and transferred to others. Through these literary vehicles, ideas emerged and were developed to the extent that thought experiment allowed. These ideas were transferred to the literature's readership, leading some to dedicate their life's efforts to the problem of spaceflight. There is a good case to be made that the first, critical technology of spaceflight was not liquid fuel rocketry, but rather the foundational technology of the book. This book is about the early origins and evolution of the core narrative underlying this process, the *spaceflight narrative*—the story that we can build machines to take us into space, travel to the planets, and explore the universe. This book is also about the origins of *space expansionism*, the belief that humanity's movement into space through spaceflight technology is a positive future that should be pursued.

It was the dedication of liquid fuel rocketry pioneers in the early decades of the twentieth century that provided the labor that enabled the ultimate realization of spaceflight—in particular, the work of Konstantin Tsiolkovsky in Russia, Robert Goddard in America, and Hermann Oberth in Germany. Each of these individuals independently laid out the fundamentals of spaceflight technology and began early physical experiments, and each one credited pre-twentieth-century spaceflight fiction as instrumental in their inspiration and motivation. In other words, each one was inspired by stories of spaceflight and believed in the core story of the spaceflight narrative enough to work to make it a reality. Before spaceflight had a technological, political, or institutional history, it had an intellectual, literary, and cultural history that spanned hundreds of years. Before spaceflight had a technological basis, it had a narrative one: a story about a future that could yet—and for many space expansionists, must inevitably—become true. As we embark upon a new era of space exploration—driven by a heady mix of geopolitical competition, conspicuous adventure tourism in space, and the space expansionist visions of billionaire moguls—revisiting the origins of spaceflight and the ideas of space expansionism can provide insight into where these forces come from and what options we may have for the paths ahead.

Today, the global culture of spaceflight includes approximately one million people around the world who produce and operate spacecraft; who work in the broader space community conducting scientific research; or who manage, promote, or otherwise support the production and operation of spacecraft and spacecraft-related components.[2] Many of these individuals, though hardly all, are inspired and motivated by spaceflight narratives produced as part of popular culture. Not included in this number, though perhaps they should be, are thousands of additional people who annually produce space-related television shows, movies, books, video games, and other space narrative content, all of whom contribute to cultural products that help support the global space industry, whether intentionally or unintentionally. As only one example, five of the top ten highest-grossing films of all time prominently propagate spaceflight themes, demonstrating that spaceflight culture, and spaceflight myths, have arguably become one of the core mainstream themes of global culture in our modern technological civilization.[3] We have spaceflight in our world today in part because we have a popular cultural narrative that continues to inspire and support its existence.

The twenty-first-century space industry is deeply intertwined with spaceflight fiction. Billionaires who were influenced deeply by spaceflight narratives in their youth simultaneously fund rocket development and finance the production of spaceflight television series that help propagate space expansionist visions of the future. Actors from popular spaceflight fiction television series travel into space as space tourists and headline space conferences. Many spaceflight professionals are also science fiction fans, with many—from spacecraft engineers to NASA administrators—professing their love of spaceflight stories and science fiction fandom. Spacecraft engineers, astrophysicists, and astronauts have credited popular spaceflight narratives—the narrative universes of *Star Trek* and *Star Wars* being two recurrent examples—as being important sources of motivation behind their interest in astronautics, astronomy, and spaceflight. NASA spacecraft projects have shared names with spacecraft from science fiction, and vice versa—sometimes due to the enthusiasm of fans outside the agency, and sometimes due to the internal fandom within NASA. It is difficult to fully understand the motivations of those who spend their careers working to achieve nearly impossible spaceflight projects, not to mention those who work toward futures in the stars that they know they will never live to see, without understanding the stories that motivate many of them to begin with.

This book examines the origins of this spaceflight narrative through its earliest historical examples, all of which were written long before any spaceflights had occurred. Today, it is more modern spaceflight narratives—such as those in twentieth- and twenty-first-century books, movies, television, and video games—that shape popular assumptions and associations related to our imagined future in space. At the origins of spaceflight, however, it was books written across the seventeenth, eighteenth, and nineteenth centuries that provided the cultural and motivation landscape for the earliest spaceflight pioneers. If we seek therefore to understand the cultural and motivational origins of spaceflight, then we need to start with these very first spaceflight narratives.

The relationship between spaceflight narratives and the spaceflight achievements that they both anticipated and motivated is one of the most enduring subjects in both science fiction and space history scholarship. This book continues in that tradition. Marjorie Hope Nicolson's pioneering *Voyages to the Moon*, Robert Lancelyn Green's brilliant *Into Other Worlds*, William Sims Bainbridge's sociological *The Spaceflight Revolution*, George Locke's bibliographic *Voyages in Space*, Ron Miller's encyclopedic *The Dream Machines*, Howard McCurdy's insightful *Spaceflight and the American Imagination*, and De Witt Douglas Kilgore's groundbreaking *Astrofuturism: Science, Race and Visions of Utopia in Space* have all explored how spaceflight fiction has evolved, and more recent analytical works have also explored how such fiction shapes (and sometimes misshapes) spacecraft, space programs, and space policy.[4] There has also been a long-standing appreciation by spaceflight popularizers that the origins of spaceflight are to be found in spaceflight stories. Many writers of popular space histories, particularly those in the mid-twentieth century—such as Frederick Ordway, Willy Ley, and Wernher von Braun—would begin their histories with early spaceflight stories, although this approach has fallen out of fashion, in part because there are now so many actual spaceflights to cover.[5]

This book builds from these earlier works, incorporates texts that were either unknown or overlooked in many of them, and focuses exclusively on the earliest period of the development of the spaceflight narrative: the centuries prior to any work by the pioneers of liquid fuel rocketry. The focus here is on the narratives that built and developed the idea of spaceflight to the point at which it was transferred to, and then acted upon by, those early pioneers. It is, in a way, a history of spaceflight focused on a time when spaceflight existed exclusively in works of literature. By investigating the earliest examples of the development of the spaceflight narrative, we can get a better appreciation for how spaceflight developed as an idea and what motivations and aspirations attended it along the way.

This book focuses primarily on the evolution of two sets of narrative vectors within the spaceflight stories that it covers: the elements related to the technical aspects of the vehicles that enable spaceflight; and the elements related to the motivations, and the economic and social context, of the individuals and groups who develop the spaceflight vehicles within the stories. Although there are a handful of works of nonfiction that seriously considered the idea of spaceflight in this early period—and it is these texts that are in some ways the most remarkable—the majority of writing about spaceflight took place, as remains in some ways the case today, in fictional stories. As such, this examination of the cultural and literary origins of spaceflight is a search for the ideas related to the technical development of, and the economic and social rationales for, spaceflight within a set of predominantly fictional narratives. In cataloguing these vectors for the first two hundred and fifty years of the conceptual development of spaceflight, this book assembles an integrated view of how people first imagined humanity traveling into space.

This book is not, however, a literary analysis or a critique of these stories, nor does it provide in-depth accounts of the broader social and historical context in which these stories were written. This book is a survey of dozens of stories across hundreds of years and as such it does not provide deep dives into the richly specific cultural contexts for each text. Many of these earlier stories are satirical and utopian narratives that were written about and against the social and economic contexts of their time, but these aspects are not central to the interests of this book. For readers interested in assessments more focused on the literary aspects of early spaceflight fiction, Marjorie Nicolson's *Voyages to the Moon* and Roger Green's *Into Other Worlds* remain unsurpassed. For readers interested in a more critical stance toward the texts of early spaceflight fiction, De Witt Douglas Kilgore's *Astrofuturism* and John Rieder's *Colonialism and the Emergence of Science Fiction* provide thoughtful critique.[6] For readers interested in assessing early spaceflight narratives within the broader context of early science fiction literature overall, Sam Moskowitz's *Explorers of the Infinite*, Brian Aldriss's *Billion Year Spree*, and Adam Roberts's second edition of *The History of Science Fiction* will provide ample comparative material.[7] The goal of this book is to lay out some of the earliest texts that contained significant thought about spaceflight and to sketch out an early intellectual history of the idea of human expansion off of the earth.

In addition to exploring the origins of the idea of spaceflight, this book will also explore the origins of space expansionism. *Space expansionism* is a term coined by Daniel Deudney in his work *Dark Skies: Space Expansionism, Planetary Geopolitics, and the Ends of Humanity*.[8] Although Deudney provides a multisentence definition of space expansionism in his book, I prefer a shorter definition: *Space expansionism* is the belief that humans both can and should travel to the planets and the stars and that they should expand human experience and presence to them. Deudney's work has been criticized within the space community, including for his negative assessment of the value of pursuing and investing in spaceflight at this time. Although I share some of these criticisms, and confess to some nontrivial space expansionist tendencies myself, it is hard to refute his description of space expansionism as a belief system or ideology or to disagree with his name for it. A very large number of the space activities that gather the most interest within the space community and the space-interested public—including asteroid mining, colonies on the Moon and Mars, interstellar travel, and encounters with extraterrestrial intelligence—are those that have not yet happened, and in some cases may never happen, and they remain much more articles of faith than lived realities. If space expansionism is a philosophy or ideology, then it is worth examining its intellectual history.

Although this book is a literary history in that the material it covers is predominantly literature and stories, its methods and approach have more in common with those of intellectual history. There are different approaches to the field, but fundamentally, *intellectual history* is "the study of intellectuals, ideas, and intellectual patterns over time."[9] The goal here is to provide an intellectual history of the origins of the ideas of spaceflight and space expansionism as they were first expressed, to provide insight into the lives of the intellectuals that expressed them, and to identify some of the patterns in their expression over time. Deudney has stated that he believes space expansionism to have developed early in the twentieth century.[10] You will see, however, that essentially all of the core beliefs and philosophical elements of space expansionism were articulated and developed in the nineteenth century or earlier. For those seeking a crash course in the more recent intellectual history of space expansionism, at least within the English-speaking world, an essentials list for reading and watching might include the following: David Lasser's *Conquest of Space* from 1931; Arthur C. Clarke's *The Exploration of Space* from 1951; Wernher von Braun's *The Mars Project* from 1953; President John F. Kennedy's address at Rice University from 1962; Gerard O'Neill's *The High Frontier: Human Colonies in Space* from 1976; Carl Sagan's *Pale Blue Dot: A Vision of the Human Future in Space* from 1994; Jeff Bezos's presentation "For the Benefit of Earth" from 2019; and Elon Musk's presentation "The Road to Making Life Multiplanetary" from 2025.[11] These are some of the core works that defined the promise of spaceflight and space expansionism during the first space race and that are defining the ideas and ideals of our current era's interest in spaceflight and space exploration. These texts, speeches, and presentations have motivated thousands of people around the world today to dedicate their lives to making a future for humanity in space a reality, and the ideas embedded in them have begun to dramatically shape our world and our beliefs about the future.[12] In many respects, however, these works can be understood as essentially elaborations of ideas and stories about the future that have their intellectual origins in texts from over a hundred years ago. By understanding the motivations and values that are embedded in the earliest stories of spaceflight we can perhaps begin to better understand the origins of space expansionism as a worldview.

The purpose here is to engage with the earliest spaceflight narratives—whether they are fiction or nonfiction, serious or satirical—as sources for an investigation into the many different ways that people first thought about the technical, economic, and social realization of spaceflight—no matter how outlandish some of these concepts may seem to us today, or indeed may have seemed to the authors themselves at the time they were written. What is perhaps most striking is that some of the very ideas that may well have been considered satirical and outlandish at the time that they were written are today remarkably close to some of the most deeply held ideas and ideals of modern spaceflight communities.

This history of early spaceflight-related literature and culture, from its origins in the 1630s to the 1890s, explores how the concept of spaceflight was developed step by step and transferred from mind to mind across the generations until it could finally be realized. After his historic orbit around the earth, astronaut John Glenn wrote, "I flew alone in Friendship 7, but thousands of brilliant, imaginative and curious minds developed the knowledge, and other thousands of skilled hands and able minds employed it, to put me there."[13] Although the majority of the brilliant, imaginative, and curious minds that enabled his flight lived in the twentieth century, a significant number of them—and the first ones—were individuals who worked to develop the concepts that led to his spaceflight by the light of candles, gaslight, and Edison's first electric bulbs.

There is an argument that we should actually consider the continuity of spaceflight's conceptual development to go much farther back than the 1630s, even all the way to the beginning of recorded time, and perhaps further. A popular starting point is Lucian of Samosata, whose "Icaromenippus" and *True History* from the second century featured, respectively, the philosopher Menippus flying to the Moon via bird wings and Greek heroes being transported to the Moon accidentally via a massive tornado. Of course, going back even farther, we could claim that the earliest human myths and legends—many of which include explanations of the movement of the Moon and the Sun—involved not only the flights of the gods but also divine vehicles for their transportation through the cosmos as well. The solar barque of Ra, *Mandjet*, which carried the god across the heavens in ancient Egyptian mythology, is one particularly evocative example. Indeed, the very earliest unambiguous physical depiction of astronomical phenomena thus far discovered—the Nebra Sky Disk, found in Germany and dated to around 1600 BC (see figure 0.1)—includes an object that has a resemblance to a boat, which some have speculated may be a representation of a solar barque.[14] That the earliest known human representation of the night sky may also include the earliest known representation of a mythic vessel that traversed it is an intriguing possibility to say the least.

The human desire to be able to travel to and live in the heavens does indeed seem timeless, and it may be one of the most fundamental of human existential ambitions. However, as appealing as these origins may

0.1 The Nebra Sky Disk, ca. 1600 BC, with the possible solar barque at the bottom of the disk. *Source:* Photograph by the author.

seem, they do not include the specific idea that matters for spaceflight: the idea that mere human beings can construct machines for travelling beyond the Earth. Spaceflight technology can be thought of as the result of the memetic transmission and evolution of that idea—the idea of humans building spaceships for space travel—through various cultural works over three hundred years until the idea was mature enough, and the overall economic and technological circumstances developed enough, to allow for spaceflight to be realized.

But why *The Spaceflight Narrative*? The word *narrative* is, of course, a synonym for *story*—and this book is fundamentally about the different stories of spaceflight wherein the idea of spaceflight itself was first developed. However, *narrative* has also come to have a broader meaning in politics, business, and cultural discourse, where the word refers to an underlying worldview, framing logic, and structure of meaning-making into which the various facts and events of the world are fit. The idea of spaceflight—particularly in the context of space expansionist thought—has in many ways evolved into a worldview about the future of humanity, and this worldview and narrative has come to structure significant aspects of our economy, politics, and indeed our daily lives. The infrastructure of the modern world—everything from our GPS-enabled devices to satellites that monitor the weather and climate to our satellite television and internet—has begun to be placed in space, in orbit around the earth. Nation-states and individuals have competed, and continue to compete, with significant expenditure of resources, both to control that infrastructure and to achieve particular stories of "firsts"—the first person in space, first person on the Moon, first person on Mars—in part in the belief that these achievements within the broader history of humanity have historic and geopolitical value.[15] All the while, we produce and consume stories about space adventures and we imagine futures—particularly when we are younger—of becoming astronauts and traveling into space ourselves. In short, we live within a world fundamentally structured by a narrative that places value on the achievement of spaceflight and that frames the long-run future of humanity as one entwined with our future among the stars. This narrative—the *spaceflight narrative*—has its origins in the first stories told about spaceflight almost four hundred years ago.

The idea of human spaceflight has its origins in the early seventeenth century, shortly after the invention of the telescope. From that moment, and through dozens of stories, the intellectual development of spaceflight began in earnest. The early intellectual development of spaceflight was propelled by numerous well-known historical figures in the United States and Europe: John Wilkins, Daniel Defoe, Cyrano de Bergerac, Washington Irving, Edgar Allan Poe, Jules Verne, John Jacob Astor IV, and H. G. Wells. But just as importantly, some of the most significant contributions to the development of spaceflight were made by fascinating characters who have been largely forgotten: Francis Godwin, David Russen, Congressman George Tucker, John Leonard Riddell, Florence Carpenter Dieudonné, Elbert Perce, Edward Everett Hale, William Leitch, and Garret P. Serviss. This book seeks to make more known their considerable accomplishments, and to explore the ways in which the idea of spaceflight evolved and was transmitted over hundreds of years and how that process motivated the ultimate realization of spaceflight itself.

In exploring these earliest spaceflight narratives, a major theme that emerges is the dominance of intrinsic motivations and private funding in pre-twentieth-century visions of spaceflight. As NASA officials and the space entrepreneurs of the twenty-first century begin to engage in more public-private partnerships for space exploration, it is important to recognize that this is not a new idea, but rather a return to one of the original concepts of space exploration expounded upon in the nineteenth century and earlier. From pre–Civil War England to Gilded Age America, not once in the first 250 years of fictional and nonfictional representations of spaceflight was it ever portrayed as a predominantly governmental affair. It was always described as the product of the efforts of individuals, small groups, or private organizations. It is not until the arrival of John Jacob Astor IV's *A Journey in Other Worlds* in 1894 that the idea that the government might support spaceflight is even hinted at, and even then, Astor was rather confident that it would be wealthy individuals like himself that would take the lead. It was only with Garret Putnam Serviss's *Edison's Conquest of Mars* in 1898 that governments were finally portrayed as taking a leading role—albeit even then only after a fictionalized Thomas Edison had privately developed a working prototype and all that was needed was for governments to scale up production. By the end of the nineteenth century, then, there began to be a more fulsome appreciation of both the economic scale required for spacecraft production and the existential national and international security implications of spaceflight capabilities, which would ultimately lead to the actual realization of spaceflight under governmental leadership in the mid-twentieth century. Prior to this, however, the pursuit of spaceflight was seen predominantly as a personal endeavor and a fulfillment of intrinsic desires.

Another major theme of this book is the extensive role that American, British, and French writers played in the early intellectual origins of spaceflight. It was predominantly British and French writers that laid the first intellectual foundations for spaceflight considerations in the eighteenth century, with there being multiple fictional stories—and, even more importantly, two nonfictional treatises by John Wilkins and David Russen—written on spaceflight concepts. It was then predominantly American intellectuals that picked up the baton and developed the concept more fully in the nineteenth century. There was no other nation in the nineteenth century whose spaceflight literature included a combination with the likes of Edgar Allan Poe, John Leonard Riddell, Edward Everett Hale, a writer with the solar-system-scale vision of the anonymous author of *Man Abroad*, John Jacob Astor IV, and Garrett Putnam Serviss. Although Russia in the nineteenth century might have attained greater philosophical and technical depth—with Nikolai Fyodorov, the Cosmist philosophers, and Konstantin Tsiolkovsky's derivation of the rocket equation in 1897—the narrative diversity and economic sophistication of nineteenth-century American spaceflight fiction has no equal. Indeed, there is no known parallel in pre-twentieth-century space history to the American intellectual effort of the 1830s, when there were at least three independent considerations of the possibility of spaceflight occurring across the country—with Edgar Allan Poe, John Leonard Riddell, and Edward Everett Hale and his fellow Harvard students. To note that this flowering of thought coincided with the Great Moon Hoax and the origins of the American observatory movement suggests that there may have been an important connection between the popular enthusiasm for astronomy in early nineteenth-century America and the alacrity with which American intellectuals applied themselves to the idea of spaceflight.

The analytical structure of this book aims to explore the long-run development of the spaceflight narrative through a mix of focused literature surveys and literary biographies across a wide span of time and geography. As the total amount of spaceflight fiction produced even prior to 1900 is so extensive, this book cannot cover all relevant works of spaceflight fiction from its origins in the 1600s until 1900, nor all the myriad themes that an analysis of these works could entail.[16] The focus here is on the technical means of spaceflight articulated in the works, as well as on the social and economic motivations for the spaceflights described. The analysis will take the form of three focused literature surveys (early spaceflight literature from 1630 until 1800, primarily English and French literature; American spaceflight literature from 1800–1850; and American spaceflight literature from 1850–1898) as well as three deep dives into iconic spaceflight narratives, with associated biographies of their writers (Jules Verne, Edward Everett Hale, and H. G. Wells and Garrett P. Serviss together). I do not claim that this is a comprehensive assessment. I have sought to be comprehensive in pre-nineteenth-century British and French spaceflight narratives, as well as the American spaceflight narratives of the nineteenth century up until Goddard was able to read *The War of the Worlds* and *Edison's Conquest of Mars* in his local newspaper in 1898. This focus—in addition to my own foreign language limitations—means that there are interesting and important spaceflight writers and thinkers that are not included—notably, Kurd Lasswitz and Hermann Ganswindt in late nineteenth-century Germany, Henry de Graffigny and Georges Le Faure in late nineteenth-century France, and the entire Russian pre-Tsiolkovsky spaceflight oeuvre. By the late nineteenth century, there are so many spaceflight narratives emerging around the world that it becomes virtually impossible to assess them all, a sign of how common a story it had already become before a single year of the twentieth century had dawned.

Chapter 1 describes the relevant English and French stories involving spacecraft from their first appearance in published literature in the 1630s through to about 1800. Particular attention is paid to Johannes Kepler's early spaceflight-related speculations and to the earliest seventeenth- and eighteenth-century spaceflight stories—notably, those of Francis Godwin, John Wilkins, and David Russen. The works of Godwin and Wilkins, despite being classics of the early spaceflight genre, remain significantly underappreciated, and the unique intellectual contributions of David Russen are almost entirely unknown. The impact of these stories on society more generally is also discussed, including the London plays that had lunar voyage themes and Daniel Defoe's later spaceflight concept in *The Consolidator*. The most significant spaceflight narratives from France are also covered, including those of Cyrano de Bergerac. The numerous female authors who contributed spaceflight narratives during this period are discussed, including Margaret Cavendish, Aphra Benn, and Marie-Anne de Roumier-Robert. The chapter describes the initial explosion of interest in lunar voyages and spacecraft that resulted from the invention of the telescope, followed by the reduction in enthusiasm for the subject after the discovery of the vacuum and its existence outside the atmosphere of Earth, a seemingly insurmountable obstacle.

In chapter 2, the focus shifts to the first American stories of spaceflight in the late eighteenth and early nineteenth centuries, with an exploration of how the combination of early Industrial Revolution technologies and the youthful enthusiasm of a new nation resulted in a unique and dramatic flowering of innovation in spaceflight concepts. The first American spaceflight narrative connects spaceflight with religious and utopian themes, while interest in the subject by Washington Irving demonstrated an early connection between space travel and conquest. The story of a voyage to the Moon by George Tucker in 1827 stands out in this context as the earliest description of a spacecraft that would seem familiar to a Mercury astronaut. The case for Edgar Allan Poe as the grandfather of realistic spaceflight fiction will also be made with an exploration of his influential work "The Unparalleled Adventure of One Hans Pfaall." To round out the chapter, we dive into the incredible story of polymath scientist, adventurer, and politician John Leonard Riddell, who gave the first known public lecture on the subject of space travel in New Orleans in 1847.

The conceptual development of spaceflight reaches an early high point in chapter 3, which dives into Jules Verne's *From the Earth to the Moon* from 1865, arguably the single most important document in the transmission and evolution of the concept of spaceflight prior to the early twentieth century. Through this story, Verne communicated a basic knowledge of the physics of spaceflight to the pioneers of liquid fuel rocketry and described with unprecedented insight the technical parameters, and the social and economic context, of space travel. The spacecraft and launch system design that Verne developed was plausible enough that when the novel was originally serialized in France, people wrote to Verne asking to be passengers on the flight. Furthermore, Verne articulates, as often only an outsider can, the precise American characteristics that made the nation such a vibrant locus for the conceptual development of spaceflight in the nineteenth century: its rapidly advancing industrial capabilities, its patriotic optimism, and the American enthusiasm for organizing private groups around special interests. Verne went even so far as to provide the first spaceflight cost estimate. Although Verne was off by two orders of magnitude from the ultimate cost of the Apollo program, he was the first to demonstrate anything even to close to an understanding of the economic scale of the project. The chapter will also explore the influence that Verne's book had on its readers in the nineteenth and early twentieth centuries. Verne's own spaceflight narrative influences—which included Poe and Godwin—will also be traced, demonstrating the direct chain of influence that extends from the earliest spaceflight concepts to our contemporary spaceflight development communities.

Edward Everett Hale's "The Brick Moon" (1869) is the focus of chapter 4, and it is in many respects the American equivalent of Jules Verne's classic tale. The book stands as the earliest text to describe both a space station and the utopian potential of space habitation—two enthusiasms that continue to motivate and drive the global space community today. Drawing from his letters and other works, the incredible life of Hale—pastor, gunrunner, philanthropic organizer, author, publisher, and chaplain to the US Senate—provides an explanatory background for the detailed descriptions that he provided of both the organizing and fundraising process for building a space station and the type of desirable life that he thought might be possible to achieve while living on one. In addition to all of this, he was also the first to describe a concept for a global navigation and positioning system—a concept that he developed in cooperation with his brother while still a teenager at Harvard. Hale was an American original and someone who deserves to be much better known as the progenitor of today's dreams of space settlement and space habitation.

The last quarter of the nineteenth century saw an increase in American spaceflight publications, and chapter 5 charts how the expanded scope and ambition of spaceflight narratives in the Gilded Age grew to encompass the entire solar system. Of particular note is the anonymous publication *Man Abroad*, which contained the first description of an American form of government taking root from the Moon to Jupiter. Millionaire moguls also started to impact spaceflight thought, with John Jacob Astor IV becoming the first member of America's wealthy elite to muse openly about spaceflight, with his romantic adventure *A Journey in Other Worlds*. American women also feature prominently in spaceflight fiction in the second half of the nineteenth century, most significantly with one of the great early writers of American speculative fiction, Florence Carpenter Diedonné, who wrote an early ode to space expansionist individualism in her fascinating *Rondah; or, Thirty-Three Years in a Star*. The narratives of space expansionism mirrored ongoing American terrestrial expansions, as well as other conflicts of the time, an example of which was *The Princess of the Moon: A Confederate Fairy Story* by Cora Matilda Ives, a member of elite Confederate social circles in Richmond during the Civil War.

The spaceflight narratives from this era in particular serve as sobering reminders that the narratives of space expansionism can contain dystopian themes as often as they do hopeful, utopian ones.

Chapter 6 explores H. G. Wells's *The War of the Worlds* and its unauthorized American sequel, *Edison's Conquest of Mars*, by American science popularizer Garrett P. Serviss. Both of these stories are notable for their frighteningly modern depictions of war between Earth and Mars, including the ultimate destruction of Martian civilization by an international military coalition of spaceships from Earth armed with disintegrators. The chapter will delve into the fascinating biographies of Wells and Serviss. Wells, in addition to being one of history's most successful science fiction writers, showed a lifelong interest in spaceflight themes and was arguably one of the first true space expansionist philosophers. Serviss was a successful American astronomer and science popularizer who had been funded by Andrew Carnegie to develop the *Urania Lectures*, a traveling lantern slide lecture series dedicated to popularizing the discoveries of late nineteenth-century astronomy. Together, these two authors and their stories directly influenced Robert Goddard to dedicate his life to the development of spaceflight technology. These stories then are a fitting end to the main body of the book as they represent the final link in the chain of influence and inspiration that connected the early narrative development of spaceflight with the beginning of the material development of spaceflight technology. With Goddard thus inspired by Wells and Serviss, and Tsiolkovsky and Oberth having already been inspired by Verne, all three of the foundational inventors of liquid fuel rocketry in America, Russia, and Germany had come to believe in the spaceflight narrative and had begun to dedicate their lives to making it a reality. From that moment in 1898 on, the spaceflight narrative had moved from a being story that was thought about or imagined solely through books and literature to being a story that people could create and live through their own lives and labors.

A few language definitions are needed. By *spaceflight,* I mean the process of traveling into space—primarily travel by humans, although not always. By *spacecraft,* I mean a physical space vehicle or robotic probe. By the *conceptual development of spaceflight,* I am referring to the cumulative knowledge process by which technical ideas for spaceflight and spacecraft are improved. By *spaceflight fiction,* I mean imagined stories that feature spaceflight in some way. By *space exploration,* I mean the process of traveling to or discovering a place in the cosmos beyond Earth. By *space development,* I mean the process of installing or improving infrastructure for the purpose of economic or social development at a place in the cosmos beyond Earth. By *space expansionism,* I mean the belief that humans both can and should travel to the planets and the stars and expand human experience and presence to them. And by *spaceflight narrative,* I am sometimes referring to a specific spaceflight story, including fictional and nonfictional ones, and sometimes to the broader metastory—the spaceflight narrative—that humans can build machines that can enable voyages into the cosmos.

A few definitions of economic terms are also needed, as I include a number of calculations that show the relative dollar values of sums from the stories in equivalent terms today. *Gross domestic product* (GDP) *ratio equivalent calculations* are used to show what a given level of expenditure would be equivalent to in our economy today if it was the same percentage of the overall economy as it was for a given year, and these calculations are adjusted by the change in the size of the overall economy, the GDP. *Production worker compensation* (PWC) *ratio equivalent calculations* are used to show what the economic scale of a given project would be today if it required the same number of implied hours of skilled worker labor as it did for a given year, and these calculations are adjusted by the change in the cost of the primary input of production for spacecraft, *skilled labor,* as measured by the PWC index. *Consumer price index* (CPI) *ratio equivalent calculations* are used to show what a given cost would be equivalent to today in terms of the comparative cost of household goods, and these calculations are adjusted by the change in the CPI. *GDP per capita equivalent calculations* are used to show what a given value would be equivalent to today relative to average per capita income, and these calculations are adjusted by the change in GDP per capita.

Throughout this work, I am interested in identifying and exploring the technical ideas for spaceflight that each spaceflight narrative includes; the social and economic motivations and dynamics related to spaceflight that are discussed; and any interesting and relevant elements of the biographies of the writers that may provide additional insight or cause for reflection. In this way, I am not seeking a grand unifying theory of spaceflight narratives, nor am I looking to critique or problematize these spaceflight narratives—though there would be no shortage of material in the spaceflight narratives covered that would be of interest to others who might be so inclined. My hope is to present, in as engaging a way as possible, a wide array of these old stories that first articulated how and why humans might travel into space, and to do so in a way that enables analysis, discussion, further research, and simple enjoyment of these fascinating works and remarkable writers.

Overall, this book seeks to explore how the stories we tell about humanity traveling into the cosmos motivate us to work to turn those stories into reality. The modern concept of spaceflight—up to and including interstellar exploration and expansion—can be understood as being the result of a multicentury, intergenerational development process, with storytelling serving as both its originating force and an enduring source of motivation. This book is an exploration of the literary origins of spaceflight because in recognizing that spaceflight has its origins in stories, we can start to see spaceflight and our future in space as phenomena that have their roots in books and culture as much as they do in science and technology. Spaceflight narratives can even be thought of as an input of production for spaceflight itself. Our space programs—along with their politics and economics—are downstream of the culture that surrounds them and are themselves cultural products. We should therefore pay attention to the stories we tell about spaceflight, to the values they encode, to the social structures they encourage, and to the ambitions that they cultivate. If we want to think about where spaceflight and our global spaceflight culture may go in the future, it may help to look back at where it began—and it began in stories told about humans building machines that could travel into space almost four hundred years ago.

1

THE DISCOVERY OF A WORLD IN THE MOON

That 'tis possible for some of our Posterity to find out a conveyance to this other World; and if there be Inhabitants there, to have Commerce with them.
—John Wilkins, *A Discourse Concerning a New World and Another Planet*, 1640[1]

In the beginning, there was the telescope. Invented in 1608 by a Dutch spyglass maker, and initially primarily intended to meet a growing market for naval intelligence-gathering tools, the following year it made its way into the hands of a forty-five-year-old Pisan philosopher living in Padua. There, Galileo Galilei set his new telescope to the stars and the planets and made a number of extraordinary discoveries that immediately and dramatically changed the world. Perhaps the most important of these was his discovery of the moons of Jupiter, which he promptly named after and dedicated to Cosimo II de' Medici. These were the first new moving objects in the heavens—the first new "wanderers" or *planetes*—to have been discovered since the beginning of recorded time. Their discovery set in motion a series of events that transformed our understanding of the nature of the universe and humanity's role within it, including the ultimate triumph of the Copernican model of the solar system. One of the other important discoveries that Galileo made—one so obvious to us now that its importance at the time is often overlooked—was his discovery that the Moon was a world, one in many ways like our own, and that this world could potentially be visited, explored, and perhaps even traded with.

The excitement generated by the circulation of Galileo's drawings of lunar mountains and dark-colored areas provocatively labeled *mare*, the Latin word for *sea*, meant that very shortly after 1610, when Galileo published his short pamphlet *Sidereus Nuncius*—Latin for *starry messenger*—people began to engage in significant thought experiments related to how they might travel there. The role of thought experiments is a critical one in achieving scientific advances. In the course of humanity's exploration of the heavens—which has evolved from an Earth-based activity of observatories and stargazing to one of launching spacecraft and ourselves off of our planet—many individuals have devoted intellectual effort to thinking about how these achievements might one day be realized, and done so at increasing levels of detail. It would be Isaac Newton, in his *Principia Mathematica* in 1687, who would conduct the first detailed thought experiment on the fundamental physics of launching an object that could orbit the earth—by imagining a cannonball being fired from the top of a mountain-peak above the Earth's atmosphere so that it would fly so far and so fast as to follow the curve of the Earth and never hit the ground.

Half a century before Newton's detailed mathematical considerations, however, an eclectic collection of writers and thinkers—including a famous astronomer, two English bishops, and a French novelist and duelist—had begun to explore not just the technical concepts for voyages to the Moon but also the motivations and socioeconomic drivers that might cause them to happen. These types of stories would become a growing part of the cultural milieu in Europe over the next two centuries, and they helped progressively develop the concept of spaceflight to a point where it began to tentatively approach the realm of the possible. The inspiration offered by these conceptual developers, and the plausibility that they conferred on spaceflight, became a key element in driving forward the effort that would eventually translate the dream of spaceflight into reality.

To fully appreciate and investigate the origins of spaceflight, we need to trace its conceptual development during its long gestation in fictional works and speculative nonfiction treatises. Although the realization of space travel would need to await the harnessing of the requisite technologies and scientific knowledge, the concept of spaceflight was nonetheless advanced and proselytized through these early works of the imagination.[2] Ideas emerged and were developed to the extent that thought experiment allowed. Over time, these ideas were in turn transferred to a wider readership, creating a wider group of people willing to devote their personal effort to considering and confronting the challenges of spaceflight. The resulting nonmarket labor supply, spawned and nurtured by this process and by individual intrinsic interest, provided the initial resources necessary to sustain the early development of spaceflight technology.

The idea of space travel has ancient roots, reaching back to myths and legends around the world. The first known Moon voyage in written literature, however, is traced to Lucian of Samosata's *True History*, written in ancient Greek in what is today modern Turkey, in the second century AD. In a parody of the voyage of Ulysses, his heroes are transported to the Moon by a fantastical whirlwind.[3] Another story by Lucian, "Icaromenippus," depicts the Cynic philosopher Menippus intentionally traveling to the Moon and beyond by improving the mythic method of human winged flight, supposedly originally invented by Daedalus, by attaching a vulture wing to one arm and an eagle wing to the other and dispensing with the fatal wax altogether.[4] The impractically of this method aside, this is the first narrative in literature of a human invention developed for the purpose of traveling to other worlds beyond Earth. Lucian also established a conceptual trend as animal-based launch systems remained the dominant concept of spaceflight for over a thousand years, well into the seventeenth century. The examples of animal-based space launch systems are numerous: the failed attempt of Kai Kaus to reach the heavens by an eagle-pulled throne in the *Shāhnāmeh*, Firdausi's eleventh century Persian epic; the horse-drawn chariot that carries the eponymous hero of Ludovico Ariosto's early sixteenth century poem *Orlando Furioso* to the Moon; and, most importantly and transformatively, the bird-borne lunar adventures in *The Man in the Moone* by the bishop of Hereford, Francis Godwin, published posthumously in 1638.[5]

One notable exemption, however, was Johannes Kepler's *Somnium*, which was written around 1611 and published posthumously in 1634. Kepler's narrative departed from the norm by having an Icelandic demon carry his central character to the lunar surface, but it nonetheless provided the first ever scientifically grounded description of the lunar surface in a narrative form.[6] As a prime candidate to be considered the first work of science fiction, it is striking that it was written in Latin, which was still the lingua franca of European culture and scientific discourse in the seventeenth century.

In the story, the unnamed narrator, effectively a substitute for Kepler himself, describes that he was reading the legends of his adopted home of Bohemia when he came across the story of Princess Libuše. As legend has it, this mythic progenitor of the Czech people foresaw a great city "whose glory shall touch the stars" and founded the city of Prague.[7] After reading this story, the narrator falls asleep and dreams of reading another book, this time written by an Icelandic wanderer named Duracotus, who as a fourteen-year-old boy ends up being stranded on Tycho Brahe's island observatory of Hven, where Kepler himself had worked for many years. Duracotus develops a love and deep knowledge of astronomy, the *divinissima scientia*, and decides to return to his homeland in Iceland five years after leaving. There he reunites with his long lost mother, who reveals to him her cosmic knowledge, including her ability to communicate with the ancient spirits of the land. She then summons one of these spirits, using twenty-one unidentified symbols, and invokes this demon to instantaneously transport her son to the wonderful land of Levania.[8] When summoned, the spirit describes Levania as follows: "Fifty thousand German miles up in the ether lies the island of Levania. The road to it from here or from it to this earth is seldom open. When it is open, it is easy for our kind, but for transporting men it is assuredly most difficult and fraught with the greatest danger to life."[9] That the father of orbital mechanics introduces the notion of traveling to the Moon via the assistance of an Icelandic spirit may seem incongruous, but it is both delightful and appropriate that Kepler managed to presage and embody Arthur C. Clarke's famous maxim: that any sufficiently advanced technology is indistinguishable from magic.

The unnamed Icelandic spirit then proceeds to give what may amount to the first description of astronaut qualifications in terms of what constituted the "right stuff" in early seventeenth century Europe: "We admit to this company nobody who is lethargic, fat, or tender. On the contrary, we choose those who spend their time in the constant practice of horsemanship or often sail to the Indies, inured to subsisting on hardtack, garlic, dried fish, and unappetizing victuals."[10] Kepler then proceeds to provide the first detailed description of the process of space travel from an experiential perspective: "In every instance the take-off hits him as a severe shock, for he is hurled just as though he had been shot aloft by gunpowder to sail over mountains and seas . . . His limbs must be arranged in such a way that his torso will not be torn away from his buttocks nor his head from his body, but the shock will be distributed among his individual limbs."[11] In addition to this strikingly prescient description, Kepler also made a number of other farsighted conceptual leaps. Writing that "after the first stage of the trip is finished, the passage becomes easier," he noted that "the bodily mass proceeds towards its destination of its own accords" and that the supporting spirits would then slow the traveler down once the Moon was reached.[12] Kepler also suggested that it would be necessary to travel on a trajectory from the earth to a point in space where the Moon and the voyagers would arrive simultaneously, in the first known description of orbital dynamics for the purpose of human space travel.

Kepler's description of the Moon itself is even more impressive: He lays out a complete translation of the art and calculations of astronomy to the lunar context, making him without question the first to formally consider the application of astronomy on another world, almost four hundred years before modern proposals for lunar observatories and telescopes. He correctly surmised the massive temperature swings that would result from the fourteen-day lunar night cycle. He described the inhabitants of the Moon—which he divides into two populations, the Subvolvans on the near side of the Moon and the Privolvans on the far side. The Subvolvans are described as enjoying looking at the Earth in their sky much as humans enjoy looking at the Moon, and Kepler accurately describes what a solar eclipse on Earth would look like to these lunar inhabitants. In addition, much as later thinkers interested in lunar exploration and development would propose, Kepler describes lunar caves as being particularly important as "these recesses are the inhabitants' principal protection from heat and cold."[13]

His speculations on extraterrestrial life are no less fascinating. He speculated that all life on the Moon would grow to a fantastic size and that lunar inhabitants were perpetual wanderers who traverse the entire Moon in large groups, following the movement of water as it moved around in accord with the lunar night cycle, traveling via their giant legs, wings, or boats. Kepler imagined an entire world in the Moon, with life and technology awaiting discovery. In addition to the text itself, Kepler also wrote extensive footnotes for his *Somnium*, and in one of them he provides one of the first links in the intergenerational chain of community that advanced the conceptual development of the spacecraft across hundreds of years. He specifically notes Lucian's *True History* as an inspiration, and credits that reading it provided "my first traces of a trip to the moon, which was my aspiration in later times."[14] With the seventeenth-century father of orbital mechanics drawing inspiration for his story of traveling to the Moon from a writer working in a remote province of the Eastern Roman Empire over a thousand years earlier, we can start to get a sense of the long chains of intellectual influence that have linked individuals across the history of the development of spaceflight.

Impressively, Kepler is not only arguably the first to have considered the genuine possibility of space travel but also unambiguously the first to have considered the possibility of space settlement. Perhaps even more remarkably, the record that we have of his thoughts regarding space settlement appear in his *Dissertatio cum Nuncio Sidereo*, a letter Kepler wrote to Galileo in response to Galileo's *Sidereus Nuncius* and that Kepler publicly published that same year in 1610. Kepler's *Dissertatio* contains one of the most far seeing and ambitious statements about the future of spaceflight that exists in the entirety of the premodern era. In one short paragraph, he lays out his belief in inhabitants of the Moon and of Jupiter, his belief that humans could one day travel to and settle these worlds, and the idea that he and Galileo could help in supporting these future endeavors through their own astronomy:

> It is not improbable, I must point out, that there are inhabitants not only on the moon but on Jupiter too or (as was delightfully remarked at a recent gathering of certain philosophers) that those areas are now being unveiled for the first time. But as soon as somebody demonstrates the art of flying, settlers from our species of man will not be lacking. Who would once have thought that the crossing of the wide ocean was calmer and safer than of the narrow Adriatic Sea, Baltic Sea, or English Channel? Given ships or sails adapted to the breezes of heaven, there will be those who will not shrink even that vast expanse. Therefore, for the sake of those who, as it were, will presently be on hand to attempt this voyage, let us establish the astronomy, Galileo, you of Jupiter, and me of the moon.[15]

With *Somnium* written only a year after *Dissertatio*, it seems quite likely that Kepler's description of the world on the Moon and its astronomy was indeed meant to serve, if not necessarily as a guide for future lunar explorers, then perhaps as a historical marker of their imagined future existence hundreds of years before their time.

Without a doubt, Kepler's *Somnium* and *Dissertatio* are some of the most remarkable texts in the history of space travel speculation, but they were in truth only the beginning of a form of conceptual development that would proliferate and expand rapidly in the decades following their publications. A little more than a decade after Kepler's pioneering publications, it would be a French writer, one who unlike Kepler is almost never linked to early spaceflight considerations, that would precede all others in his brief but powerful description of how humanity might ultimately venture to the Moon through technological means.

Charles Sorel was one of the most important novelists of the French baroque, and his masterpiece, *Francion*, surprisingly includes not only the first mention of "engines" that could power humans to the Moon but also the first plausible sociological and economic motivations for how and why such machines might be constructed. Born to minor courtiers in Paris around 1600, he was arguably the father of the modern French novel and reportedly lived a life that was "typical of the many prudent, hard-working bourgeois and minor robe officials of the first half of the seventeenth century."[16] However, his imagination was second to none, and his masterwork, *Histoire comique de Francion*—written in his early twenties during his association with the emerging Libertine philosophical movement—contains one of the first descriptions in literature of the potential creation of a spacecraft amid the book's social satire, inventive descriptions of foreign lands, and rather copious erotic adventures and misadventures.

Sorel introduces the concept of the plurality of worlds through a fictional philosopher named Hortensius, who states: "You know that some Wise men have affirmed, that there are many worlds, which some have placed in the Planets, and others in the fixed Starrs; For my part, I believe that there is a world in the Moon."[17] Hortensius goes on to speculate about life on the Moon:

> There will be a Prince as ambitious as *Alexander*, who shall come to conquer this world; He shall make great provision of Engins, either to descend, or to ascend (for to speak the truth, I do not yet know whether we are above that world, or beneath). He shall bring with him another *Archimides*, who shall make him all

manner of structures, and ladders, by the means whereof he shall enter into the Epycicle of the Moon excentrick to our Earth, and there it shall be where he shall find some place habitable, where shall be Nations unknown, whom he shall overcome.[18]

Here, in an extended libertine novel, a philosopher describes a prince of the Moon using engines to travel to and conquer another world, accompanied by a great inventor who could make other devices to continue his conquest of the solar system. Already in the 1620s, wealthy patrons desiring conquest, and clever inventors who could aid them, were considered common aspects of life not just on Earth, but on other worlds as well, and these patrons and inventors were already imagined to be drivers of the development of engines to travel between them.

Although both Kepler and Sorel are important preludes, it is Francis Godwin's *The Man in the Moone*, published posthumously in 1638, that deserves credit for being the first attempt to detail the process of human invention for a specific technology to enable a space voyage. In so doing, Godwin also created the first space travel narrative with significant and enduring cultural impact, both in his own time and throughout centuries after. In the story, Domingo Gonsales, a Spanish aristocrat and adventurer, becomes ill while sailing home after making his fortune in the East Indies and is set on the remote island of St. Helena to recover. While leisurely awaiting his passage home, he begins to tame and experiment with the birds on the island—swan-like creatures called *gansas*—using them first to carry messages between himself and his servant who lives on the other side of island.[19] Over time, it occurs to him that he might use the gansas to fly heavier objects—and ultimately, even himself. Godwin describes his painstaking experimentation with the gansas and his eventual development of a system of ropes, weights, and pulleys, which he calls his "engine" that allows a flock of twenty-five birds working in unison to carry him into the sky (figure 1.1). He arrives at the Moon not by intention but because, as he discovers, the gansas travel to the Moon regularly as part of their annual migration. Godwin's gansa engine story is, arguably, the first known description of a spacecraft development program, albeit a largely accidental one. It is also a narrative that set the pattern of the spaceflight genre from hundreds of years to come, until roughly Jules Verne's *From the Earth to the Moon* in 1865: the story of a lone inventor

1.1 The engine in Francis Godwin's *The Man in the Moone. Source:* Francis Godwin, *The Man in the Moone: Or, A Discourse of a Voyage Thither* (London, 1638), frontispiece.

with sufficient leisure and access to materials that he could, through his personal effort and ingenuity, engage in technical experimentation, trial, error, and test flights before ultimately developing a device that could travel into space.

However fanciful and naive we may now consider Godwin's story, he attempted to incorporate some of the modern science of the day. Godwin was aware of the Copernican theory and, more than fifty years before Newton's *Principia Mathematica,* he incorporated a crude concept of gravity into his narrative. As the gansas fly upward, Gonsales notices that "the Lines slacked; neither I, nor the Engine mov'd at all, but abode still as having no manner of weight."[20] This weightlessness was described by Godwin as the result of being farther from the "secret property of the Globe of the Earth," an attractive force, that draws objects to it and gives them the feeling of weight. Once he was "free from the attractive Beames of that tyrannous Loadstone the earth," the flight of Gonsales's gansa engine to the Moon required almost no further effort.[21]

Although Godwin's story may not seem like a major breakthrough, it was the first space travel narrative to enjoy wide circulation and popularity. Godwin's story would be printed in four languages and in twenty-five editions from 1638 to 1768, and it even inspired a play by one of the first professional female playwrights, Aphra Behn, who wrote a play entitled *The Emperor of the Moon*, published in 1688. Godwin would remain a prime influence well into the nineteenth century as well, with his reach eventually extending to the other "new world": the Americas. In the original appendix to Edgar Allan Poe's epochal space travel narrative "Hans Pfaall," in 1840, Poe acknowledged only one previous work of spaceflight fiction, that of Godwin—which Poe described as "a singular and somewhat ingenious little book."[22] Arguably, however, Godwin's most significant impact was on the thinking of his contemporary, John Wilkins, a founding figure of the Royal Society and the writer of the first nonfiction treatise on the topic of space travel.[23]

In 1638, at the age of twenty-four, John Wilkins published his first nonfiction work, *The Discovery of a World in the Moone*, which discussed the physical nature of the Moon and defended the notion that it was a world like the Earth and likely had inhabitants.[24] As with Godwin, this interest in the Moon as a world was the result of the explorations of the lunar surface that had become possible following the invention of the telescope. The famed playwright Ben Jonson captured the popular enthusiasm and quick speculation that attended these discoveries in a masque that he wrote for the court of King James I in 1620:

HERALD 2: Certain and sure news.

HERALD 1: Of a new world.

HERALD 2: And new creatures in that world.

HERALD 1: In the orb of the moon.
HERALD 2: Which is now found to be an earth inhabited.
HERALD 1: With navigable seas and rivers.
HERALD 2: Variety of nations, policies, laws.
HERALD 1: With havens in't, castles, and port-towns.
HERALD 2: Inland cities, boroughs, hamlets, fairs, and markets.[25]

In his treatise, the young Wilkins argued for thirteen propositions in support of his view that the Moon could be another Earth-like world. Proposition 2 asserted "that a plurality of worlds doth not contradict any principle of reason or faith," while Proposition 13 claimed "that tis probable there may be inhabitants in this other World, but of what kinde they are is uncertaine."[26] Although his 1638 treatise contained some speculation on the possibility of space travel, it was only after reading Godwin's work, which had been published anonymously in the same year, that Wilkins added a groundbreaking fourteenth proposition in his *A Discourse Concerning a New World & Another Planet* in 1640. This proposition boldly stated: "That 'tis possible for some of our Posterity to find out a conveyance to this other World, and if there be Inhabitants there, to have Commerce with them."[27]

Wilkins's proposition of an Earth-to-Moon conveyance took inspiration from Godwin. Wilkins advanced a similar notion of gravity as a sort of "Magnetical Vigor" that decreased in strength with distance from the Earth. On this basis, he postulated that "if a Man could but Fly, or by any other means get Twenty Miles upwards, it [would be] possible for him to reach unto the Moon."[28] He also treated seriously Godwin's notion of bird migration, asserting that "by this means, 'tis easily Conceiveable how once every Year a Man might Finish such a Voyage; going along with these Birds at the beginning of Winter, and again Returning with them at the Spring."[29] Wilkins's interest in Godwin's idea of the lunar migration of birds likely contributed to its prominent treatment in the first treatise on bird migration in England, written by Charles Morton—a student of Wilkins at Wadham College, Oxford. Morton calculated that the round trip to the Moon would take four months for the birds, who he imagined traveling there and back at 125 miles per hour while sleeping and living off their stored fat.[30]

As amusing as these seventeenth-century lunar goose migration speculations may be, they cannot take away from the fact that it was nonetheless John Wilkins who was the first to openly declare in a nonfiction treatise the eventual likelihood of spaceflight, stating even in his first edition of the work: "I do seriously, and upon good Grounds, affirm it possible to make a Flying Chariot. In which a Man may sit, and give such Motion unto it, as Shall convey him through the Air. And this perhaps might be made large enough to carry divers Men at the same time, together with Food for their *Viaticum*, and Commodities for Traffique."[31]

He did not speculate on how such a flying chariot might be constructed but nonetheless asserted that it would likely eventually become possible based on its evident physical possibility and that "the Industry of Future Times, Assisted with the Labours of their Fore-Fathers, may reach that Height which we could not Attain to."[32] Wilkins even laid out what he saw as the three principal technical challenges of traveling to the Moon, which, if generously interpreted, still stand as a reasonable rough summary of the nature of the problem:

1. The First, taken from the Natural Heaviness of a Mans Body, whereby it is made Unfit for the Motion of Ascent, together with the Vast Distance of that Place from us.
2. From the Extream Coldness of the Aethereal Air.
3. The Extream Thinness of it.[33]

The young Wilkins was confident, however, that human invention would surmount these challenges: "So that notwithstanding all these seeming impossibilities, tis likely enough, that there may be a means invented of Journying to the Moon; And how happy shall they be, that are first successful in this attempt?"[34] Wilkins did not only foresee that a journey to the Moon would be possible; he foresaw at least a glimmer of the joyous celebration that would arise across the globe when indeed it was finally achieved.

There is a good case to be made that John Wilkins should be considered the first space expansionist philosopher. Not only did he reason his way into believing that it would be possible to build vehicles that could voyage to the Moon, but also he proposed that there would be colonies on the Moon once the art of flying was discovered, and he waxed eloquent, in language that foreshadowed Carl Sagan by three hundred years, on his belief that "a mansion" in the "wider spaces above" was a central feature of humankind's destiny:

> What great matter is it to be a Monarch of a small part of a Point? Might not the Ants as well divid a little Mole-hill into divers Provinces, and keep as great a stir in disposing of their government? . . . All this place wherin we warre, and travell, and dispose of Kingdoms, is but a point far lesse than any of those small stars, that at this distance are scarce discernable. Which when the Soul does seriously meditate upon, it will begin to despise the narrownesse of it's present Habitation, and thinke of providing for it selfe a mansion in those wider spaces above, such as may bee more agreeable to the noblenesse and divinity of it's Nature.[35]

Here then is perhaps the first articulation of a motivation for space settlement. Kepler had assumed that there would someday be interplanetary settlers, but we have little information to suggest why he thought such settlers might set out for new worlds in the vastness of space. John Wilkins, however, who would later in his life become the bishop of Chester, implies that he sees a potential future for humanity in space as part of a natural extension of the "noblenesse and divinity" of the nature of the human "Soul." A fundamentally religious belief is therefore present at the intellectual origins of spaceflight. There is an argument to be made that this faith-based approach to spaceflight, and to discussions of potential human futures in space in particular, continues very much to the present day.

Wilkins goes further, too, and begins to suggest the types of people who might ultimately engage in spaceflight and venture to the Moon. He considers the basic logistics of a lunar expedition, estimating that the distance will be "about 179,712" miles—relatively close to the actual average of around 238,855 miles—and determines that if a person could travel at the speed of the fastest bird and could fly a thousand miles a day, it would still take him upward of half a year to get to the Moon. This presents a problem as he would have to carry all his supplies with him, and he would find no resupply stations along the way: "I believe he shall scarce find any lodging by the way. No inns to entertain passengers, nor any castles in the air (unless they be enchanted ones) to receive poor pilgrims, or knight errants."[36] Wilkins lists this lack of lodging as a significant issue, one that might seem to preclude such a voyage entirely, but he nonetheless maintains a belief that a voyage to the Moon might someday nonetheless be possible. Interestingly enough, humanity would eventually build multiple "castles in the air"—from Skylab to the Mir Space Station to the International Space Station—and NASA and its international partners have proposed building the first lodging by the way of the Moon in the form of the Lunar Gateway, a small space station planned to orbit the Moon as part of the Artemis program. From a certain point of view, the Apollo astronauts shared much in common with the knights-errant of medieval lore, both in their romantic wanderings and their (supposedly) chivalric virtues. As for poor pilgrims making their way to the Moon, this is a theme that we will encounter again. For now, we might simply note that Buzz Aldrin, the second person to walk on the lunar service, made time to take communion while on the lunar surface, and he did so while earning the standard salary of a colonel in United States Air Force.

Within thirty years of the invention of the telescope and the discovery of the planetary nature of the Moon, then, a young man in his mid-twenties had described seriously the technical challenges of spaceflight, concluded that they would likely be solved, envisioned human migration into space as part of nothing less than the destiny of the human soul, and foresaw a future in which adventurers and pilgrims might want for castles in the air on their way to the Moon. The philosophical consideration of humanity's expansion into the vastness of the cosmos had thus begun. That the origins of serious spaceflight consideration are to be found in the writings of not just one but two different Anglican bishops is something that the Anglican church might want to consider politely taking a little more credit for. Regardless, the contributions of Francis Godwin, the seventy-third bishop of Hereford, and John Wilkins, the fourteenth bishop of Chester, to the development of humanity's movement into the heavens are nonetheless worth celebrating.

From Godwin and Wilkins on, lunar travel narratives became increasingly mainstream. The Moon had now been incorporated into the mental landscape of potential human exploration, as indicated by its inclusion in Peter Heylyn's *Cosmographie in Four Bookes, Contayning the Chorographie and Historie of the Whole World,* first published in 1652. In Heylyn's *Cosmographie,* the Moon was presented alongside other new worlds, such as the Americas, with the following description:

> The New World in the Moon was first of Lucian's discovering. . . . But of late times, that world which he there fancied and proposed but as a fancy only, is

become a matter of more serious Debate; and some have laboured with great pains to make it probable that there is another *world* in the Moon, inhabited as this is by persons of divers Languages, Customs, Politics, and Religions; and more then so, some means and ways proposed to Consideration for maintaining an entercourse and Comerce betwixt that and this.[37]

This burgeoning enthusiasm for Earth-Moon commerce aside, major new advances in the conceptual development of methods to explore this new world in the Moon would be relatively few and far between until the nineteenth century. By the end of the seventeenth century, new discoveries in the nature of the vacuum of space, gravity, and the vast distances between the planets made the Moon and spaceflight seem, in some ways, more remote than it had in 1640.[38] It would take the advances of the Industrial Revolution to make the Moon again seem plausibly within human reach and worthy of a redoubling of intellectual effort.

There were, nonetheless, several noteworthy contributions to the concept of spaceflight in the intermediate period. Perhaps the most well-known of these are the ones from Savinien de Cyrano de Bergerac, published posthumously as *L'Histoire des États et Empires de la Lune* and *L'Histoire des États et Empires du Soleil* in 1657 and 1662. Cyrano de Bergerac's first-person narrative describes a multiplicity of different experiments for achieving his voyage to the Moon, and later to the Sun. His first attempt is through a system of vials filled with Sun-attracted morning dew, which allows him to fly into the sky, though not all the way to Moon, before landing in the French colonies of "Nouvelle France,"across the Atlantic in what today is Canada. It is there, in the central square of *la ville de Quebec*, that rockets make their first appearance in a spaceflight narrative. After discussing the plurality of worlds hypothesis and the infinite nature of the universe with the Viceroy of New France, Cyrano returns to his quest for the Moon and builds a new machine with spring-based wings, although he fails to achieve flight with it, bruising himself in the process. He repairs to his lodgings to attend to his injuries by covering himself with beef marrow and then returns to his machine only to find it missing. He eventually finds it in the marketplace in Quebec where, as part of the celebration of the feast day of Saint-Jean-Baptiste, the Quebecois revelers had attached fireworks to Cyrano's flying machine. Cyrano tries to stop the fireworks from being lit but fails, and instead ends up in the machine and is launched

1.2 Cyrano de Bergerac's rocket launch from Quebec. *Source:* Cyrano de Bergerac, *The Comical History of the States and Empires of the World of the Moon* (London, 1687), illustration.

into space. It is therefore celebratory *Canadiens*, rather than Cyrano himself, who build the first spaceflight rocket system in literature, and do so with some attention to technical detail: "For so soon as the Flame had devoured one tier of Squibs, which were ranked by six and six, by means of a Train that reached every half-dozen, another tier went off, and then another."[39] Dozens of sequenced rockets thus provide the first stage of Cyrano's launch system, while the second stage—the aforementioned body grease of animal bone marrow—provides the final impulse through the supposed force of the sucking of bone marrow by the Moon when in its waning phase (figure 1.2).[40] Although these works are often dismissed as pure satire, Cyrano de Bergerac—an intellectual and an ex-soldier of aristocratic heritage—was well educated in astronomy, having been a student of the empirical scientist and astronomer Pierre Gassendi. Cyrano de Bergerac himself was an advocate of the Copernican system, the habitability of the planets, the existence of planets around other stars, and the theory of an infinite universe.[41] Self-admittedly influenced by Godwin's work, which was translated into French in 1648, Cyrano de Bergerac described an incredibly innovative and energetic spaceflight experimentation process that includes the first description of rocket-based spaceflight in *L'Histoire des États et Empires de la Lune*, published exactly three hundred years before the flight of *Sputnik 1*.

Other notable forays into space-travel-related considerations in the seventeenth century include two by prominent female authors. The first known was Margaret Cavendish, the duchess of Newcastle. Cavendish had been a lady-in-waiting to Queen Henrietta Maria (wife of Charles I), was a prolific writer, and was the first woman to have been inducted into the Royal Society (albeit posthumously, over two hundred years after her death). Cavendish's groundbreaking, fantastical tale *The Blazing World*, from 1666, does not include space travel per se—as the titular Blazing World is reached simply by traveling to the North Pole, where the two worlds are described as being connected—but it is another example of a text sometimes considered the first work of science fiction. It features arguably the first world-building of a truly alien world, one full of birdmen and other human varieties of terrestrial animals, over which the female protagonist ultimately desires to rule as empress.[42] Indeed, Margaret Cavendish even goes so far as to insert herself into the story with the new empress inquiring whether another world might conquered so that Cavendish could rule over it: "The spirits answered, that there were more numerous worlds than the stars which appeared in these three mentioned worlds. Then the Empress asked, whether it was not possible, that her dearest friend the Duchess of Newcastle, might be Empress of one of them? Although there be numerous, nay, infinite worlds, answered the spirits, yet none is without government. But is none of these worlds so weak, said she, that it may be surprised or conquered?"[43] The idea of planetary conquest, thus introduced, would grow in popularity until it featured prominently as one of the most common themes of spaceflight narratives.

Aphra Behn, who was one of the first professional female playwrights, was arguably the first ever to deal with the topic of space-obsessive behaviors in her play *The Emperor of the Moon* in 1687. The play was a farcical tale of young women and their suitors, and it features a father so obsessed with the Moon that he will not let his daughters consort with mere humans, preferring them instead to marry men from the Moon. The father is the first—but hardly the last—true lunar obsessive in literature, and he is presented as a person who "so Religiously believes there is a World there, that he discourses as gravely of the People, their Government, Institutions, Laws, Manners, Religion and Constitution, as if he had been bred a *Machiavel* there."[44] When asked how her father became so besotted, the protagonist Elaria responds by mentioning the increasing profusion of lunar narratives, including those of Lucian, Godwin, and Wilkins: "With reading foolish Books, *Lucian's Dialogue of Icaromenippus*, who flew up to the Moon, and thence to Heaven; an Heroick business called, *The Man in the Moon*, if you'll believe a *Spaniard*, who was carried thither, upon an Engine drawn by wild Geese; with another Philosophical Piece, *A Discourse of the World in the Moon*; with a thousand other ridiculous Volumes too hard to name."[45] By the 1680s, then, the idea of traveling to the Moon had become so commonplace that a popular play would be written that featured interplanetary obsession as a core part of the plot.

Not all interesting spaceflight narratives from this era would become popular and well recognized, however. One of the least-known major eighteenth-century achievements in the intellectual history of spaceflight's conceptual development was achieved by a Kentish schoolteacher, David Russen, who in 1703 published one of the most interesting and underappreciated early texts of spaceflight consideration. Russen's *Iter Lunare* is a nonfiction response to Cyrano de Bergerac's stories, with Russen dutifully outlining Cyrano de Bergerac's views on the challenges of traveling to the Moon and then speculating beyond them. In so doing, he describes a method of reaching the Moon that was unmatched in its monumental scale until Jules Verne's massive cannon in *From the Earth to the Moon*. As a serious, nonfiction treatment of the concept of spaceflight, *Iter Lunare* is preceded only by Wilkins's *Discourse*, and it shows a number of significant refinements in thinking.

Russen first outlines what he sees as the principal challenges: "Two things seem to impede our journey thither, the propensity of all Earthly Bodies composed of Matter to tend downwards, which causeth in them an Ineptitude to ascend, and the Medium through which they are to pass, which is unfit for Animals of this Earth to breathe in. As for the distance, it is no essential hindrance, since, were it farther, it might be surmounted with Labour, and time proportionably allowed."[46] Like Wilkins, and directly building from his work, Russen explicitly laid out the assumptions that underpinned his considerations:

> That the Atmosphere of Air which encompasseth our Globe, is at most not above 20 miles from the Superficies of the Earth.
> That the higher we Ascend, the less weight we have.
> That the *Aura Aetherea*, is pure, thin and rarified.
> That in it we have no Gravity, and consequently no propensity of descending.
> That the Moon is encompassed, with an Orb of Vaporous Air like the Earth.
> That when we arrive within the Sphere of the Moons activity, we shall as naturally descend to it, as now we do towards the Earth.
> These are the several Regions through which we are to pass, what carriages we shall have is our Enquiry.[47]

It does not seem that Russen was familiar with Newton or his concept of gravity—he refers only to the "magnetical attraction of the Earth"—or with the notion that interplanetary space might be a vacuum—he uses the term *Aura Aetherea* to refer to the region between the Earth and Moon and posits that it could potentially consist of fire or water. Nonetheless, his consideration of the difficulties of spaceflight is more thorough than Wilkins's. He recognized the need to provide breathable air from the Earth for the journey, that provisions for food and sleep would be required, and that the mass of these would increase the force needed to reach the Moon: "Yet the more we take with us to prevent Famine, the more Luggage we shall have to encomber us, and this requires a greater carriage, all which will encrease the weight, and make the ascent the more difficult, since the least crum of Provision will not ascend without some Extrinsick force to drive it forwards."[48] In considering the challenge of provisions, Russen recognized that the ability to reprovision in space would simplify the logistics of the journey. In doing so, he was the first to imply the potential future development of a way station on the way to the Moon—and, perhaps just as importantly, the first drinking establishment in space—when he lamented that "'tis to be supposed no Inns, Ale-houses, or Places of entertainment, are as yet built in that Pathless way to the Moon."[49] This sentence is clearly derived from the nearly identical one found in Wilkins's work, but Russen adds a provocative "as yet" and suggests that while there are likely no way stations on the way to the Moon, such way stations could nonetheless someday potentially be built. For this provocative eighteenth-century suggestion of the potential development of future cislunar infrastructure, Russen deserves a more prominent place in the spaceflight canon.

Russen evaluated the ability of the concepts proposed by Godwin, Wilkins, and Cyrano de Bergerac to meet the challenges of spaceflight and largely determined them to be "vain and ineffectual."[50] He dismissed the notion of ascent by spirits, demons, or the types of intrinsic attraction in Cyrano de Bergerac's *Histoire Comique* and insisted: "Whatever means of Ascent can be invented, they must be violent and praeternatural, artificial, and done by some kind of Mechanism, either vital of Birds, or Inanimate, yet Forcible by Springs or Instruments; which whether any can effect, is a matter of Controversie."[51] Of the mechanistic, artificial methods that had been proposed, he determined that the method of adding wings to the arms would be ineffectual as the labor required would be excessively tiring: "I grant by such means Men may become lighter, the Feathers and Wings buoying them somewhat upwards, but it will never reach so high as the Moon."[52] He noted Wilkins's concept of the flying chariot, but thought it implausible based on the fact that "the Matter of which it must be framed, will be naturally heavy, and apt to descend; That the Principle of Motion, or Ascent upwards, must be forcible, both to carry up the Engine, and the materials laid in it; That this motion must be caused by Springs, Wheels or Weights, and those must be complicated within the Body of the Engine."[53] Russen concluded that that "the forcible Ascent" of any flying chariot "will cease, when the Spring which raised it hath spent its Strength; and so the whole will again descend, except we could discover a Perpetual Motion," a discovery which he did not think was likely.[54] In fact, Russen thought Godwin's method of harnessing the gansas the most plausible, but noted that it required an actual lunar migration of birds, when "most probable Conjectures conclude, they only seek some warmer Climate farther Southwards."[55]

With these previously conceived methods largely rejected, Russen then described his own concept for reaching the Moon with a device whose immense scale far surpassed anything that had previously been considered for space travel:

> Since Springiness is a Cause of forcible motion; and a Spring will, when bended and let loose, extend it self to its length; could a Spring of well-tempered Steel be framed, whose Basis being fastned to the Earth, and on the other end placed

a Frame or Seat, wherein a Man, with other necessaries, could abide with safety, this Spring being with Cords, Pullies, or other Engins bent, and then let loose by degrees by those who manage the Pullies, the other end would reach the Moon where the Person who ascended landing, might continue there, and according to a time appointed, might again enter into his Seat, and with Pullies the Engine may again be bent, till the end touching the Earth, should discharge the Passenger again in safety.[56]

Russen's description is of a device of unprecedented size—a spring that would stretch the entire distance from the Earth to the Moon. To limit this distance, Russen proposed that the gigantic structure could be built and operated in the high mountains of Asia.[57] Although he does not provide an estimate for this distance, Russen assumed the limit of the Earth's atmosphere to be twenty miles, and he would have read Wilkins's estimate of a distance to the Moon of almost 180,000 miles. Russen recognized the incredible scale of the project he was proposing, but he nonetheless asserted that "There is Mettal enough in the World of Iron or Steel to compose it, Men enough may be had to frame it, Firing enough to work it, and Money enough to defray the Charges."[58] The complexity of the endeavor did, however, mean that Russen thought it unlikely to be accomplished—not because it was physically impossible but because the "Covetousness, Vice, Intemperance, Slothfulness and Ignorance" of mankind meant that those with the desire to build such a device would be unable to do so "'till *Vulcan* and his *Cyclops* come among us to undertake it."[59] Russen had seriously evaluated the challenges of traveling to the Moon, had conceived of a truly massive machine for overcoming them that he believed it was possible to build, and had determined that the real impediment was neither technical nor economic, but cultural.

Russen thus earnestly grappled with the challenges of spaceflight in his eighteenth-century mind and ultimately concluded that humankind was not ready for the journey: "That while we dwell in these Tabernacles of Clay, we must content our selves with this Earth [God] hath allotted us, reserving the farther Enquiry hereof till a future Estate, when more of the Infinity of God's Works shall be discovered to our Understandings."[60] Rather than the youthful optimism of Wilkins, Russen brings a thoughtful skepticism to his consideration of space travel. Concluding that he will not see flight and space travel in his lifetime, like many future spaceflight enthusiasts and thinkers in the centuries to come, he contents himself with the hope that he has "framed a Foundation, on which others may build."[61] Although *Iter Lunare* never reached the popularity and influence of Godwin's work, with only two printings in 1703 and 1707, it is indicative of the significant accumulation and transfer of knowledge on the subject that had occurred since Godwin. That Russen hoped that his work would be built on by others in the future provides further evidence, in addition to the writings of Kepler and Wilkins, that the concept of an intergenerational community of spaceflight development was beginning to take root over three hundred years ago. That an eighteenth-century schoolteacher in Hythe, Kent, conceived of the first Apollo-scale project for reaching the Moon is a testament to the intrinsic interests driving spaceflight thought experiments in both fiction and nonfiction across successive generations.

Throughout the eighteenth century, much of the pessimism toward near-term spaceflight seen in *Iter Lunare* continued, with the result that there were relatively few significant new conceptual advances. The pessimism reached a particular apex in the anonymous *A New Journey to the World in the Moon,* published in 1741, which derided the ideas of Godwin and Wilkins and explicitly stated that "no material Engines, nor any possible Inventions can ever convey our Bodies to the World in the Moon."[62] The work therefore has its narrator travel to the Moon merely by "a Spiritual Analogy" and thinking at the Moon intensely from atop a high mountain.[63] The majority of stories during this time include voyages to the Moon primarily for the purpose of general philosophizing or of satirizing more terrestrial topics. Many of the spaceflight concepts and considerations included are therefore incidental, but some are nonetheless interesting and noteworthy. One such example is the highly satirical *A Voyage to Cacklogallinia,* published in 1727 by a pseudonymous Captain Samuel Brunt. This work interestingly provided the first narrative of a public joint stock company being formed for the purpose of returning lunar materials to Earth—in this case, gold—in the first explicit description of a Moon-mining venture in fiction. However, the project is undertaken predominantly not by a group of industrious men and women, but by a nation of bird people who fly themselves to the Moon under their own power, carrying the pseudonymous Captain Brunt with them in a flying palanquin.[64]

A Voyage to Cacklogallinia does, however, include several fascinating elements that presage those experienced by some private-sector space ventures on the public markets almost three hundred years later. When the joint stock company is created, shares are sold and start to skyrocket, and fortunes begin to be made. The bird person responsible for managing the project's preparations, however, is called Volatilio, likely a tongue-in-cheek comment on the volatility of share prices. Test flights with the palanquin are undertaken on a hilltop to see how the palanquin, and Captain Brunt inside it, manage with the coldness and the thinness of the air, with the later mitigated by the use of breathing sponges. With every hint of success during the test flights, the company shares receive a boost. However, one unsuccessful test flight by Volatilio causes the share price to collapse, at which point the king of the bird people and his courtiers insist that Captain Brunt reassure markets and proceed promptly with the voyage, which he does—traveling to the Moon for further satirical adventures, including discussions with giants and the souls of formerly earthly inhabitants. Although share price volatility is far from specific to space ventures, it is nonetheless impressive that an anonymous writer in the early eighteenth century foresaw the way in which successful and unsuccessful flight tests in a spacecraft development effort might influence the stock prices of new space ventures.

A work with greater significance to the development of the spacecraft, however, comes from Daniel Defoe. The writer of *Robinson Crusoe*, and thereby the inventor of the *robinsonade* genre of fiction that contributes much to marooned spaceflight narratives up to the present day (Andy Weir's *The Martian* being only one example), in fact also contributed to the conceptual development of the spacecraft in another story, similar to that of Charles Sorel, that features the imagined spaceflight technology of extraterrestrials. In his *The Consolidator* in 1705, Defoe, who was a student of John Wilkins's protégé Charles Morton, described in detail an "Engine" for traveling to the Moon, which was the first to explicitly make use of a consumable fuel:

> But above all his Inventions for making this Voyage, I saw none more pleasant or profitable, than a certain Engine formed in the shape of a Chariot, on the Backs of two vast Bodies with extended Wings, which spread about 50 Yards in Breadth, compos'd of Feathers so nicely put together, that no Air could pass;

and as the Bodies were made of *Lunar Earth* which would bear the Fire, the Cavities were fill'd with an Ambient Flame, which fed on a certain Spirit deposited in a proper quantity, to last out the Voyage; and this Fire so order'd as to move about such Springs and Wheels as kept the Wings in a most exact and regular Motion, always ascendant.[65]

The inventor of this impressive machine is not a human but rather a lunar citizen, one described as having landed in China, to whom Defoe attributes the many technical advances of Chinese civilization.[66] Nonetheless, Marjorie Nicolson noted that Defoe "alone, among scientists and men of letters, anticipated the gasoline age" in describing how an "Ambient Flame" and a burning "Spirit" would power the engine.[67] Although it was not yet a rocket engine, Defoe deserves credit for anticipating some of the elements of the engine that would ultimately enable space travel. Von Braun's V2 engines would ultimately also operate on a spirit—grain alcohol—mixed with water and combined with liquid oxygen for combustion. Defoe's engine, which can contain and harness "sublime Flame" in "Receptacles," which then "vent at the Cavities appointed" and thereby provide "Life and Motion" to the vehicle, is described remarkably similarly to the functioning of a modern rocket engine—provided that we can overlook the fact the engine is built significantly from lunar-derived feathers.[68] There is even discussion of the risk that the resulting fire presents to the composite feather, an early consideration of the challenges that would be posed to various materials by the high combustion temperatures required for spaceflight. The advances in spaceflight conceptual development seen in Defoe's work are impressive, and with an explicit reference to Wilkins in his text, he represents another link in the multigenerational chain of influence and inspiration that has led to the development of the spacecraft.[69]

The Consolidator also represents one of the earliest narratives that explicitly engages spaceflight with broader cultural themes around colonization and that introduces for the first time the idea of extraterrestrial intelligences improving life on Earth. Defoe notably begins the book by praising "the Czar of Muscovy" and noting particularly how he "built Forts, planted Collonies and Garisons at proper Distances; where though they are seated in Countries intirely Barren, and among uninhabited Rocks and Sands; yet, by his continual furnishing them from his own Stores, the

Merchants travelling are reliev'd on good Terms, and meet both with Convoy and Refreshment."[70] Adjust the language for modern terms and one could easily imagine the same sentiment expressed today by advocates of space settlement and space expansionism. After singing the praises next of the "Ancient, Wise, Polite, and most Ingenious" Chinese, Defoe's narrator describes the engine in the "Chamber of the Emperor's Rarities," and goes on to imply that the significant technological advances in China were, at least in part, attributable to a man who "was no Native of this World, but was Born in the *Moon*" and who traveled to Earth to explore it before being persuaded by the Chinese emperor to "stay and improve his subjects," with the result that it was "no wonder the *Chinese* are such exquisite Artists, and Masters of such sublime Knowledge."[71] He goes on to describe a particular Chinese monarch that developed a "desperate and implacable Resolution" to "fly up to the Moon" and sought out and funded all manner of inventions that "were so preposterous and ridiculous that his subjects seeing him going on to ruin himself, and by Consequences them too, unanimously took arms," believing their prince to have become a "Madman."[72] Even in the early eighteenth century, then, it was already possible to imagine negative popular opinion dooming a spaceflight enterprise, even one aided by extraterrestrials.

Defoe was also prescient in describing the potentially massive social and economic impact of space travel, noting that "it can no more be a Wonder, if exalted in the Center of this famous Engine, a whole Nation should be carried up to the World in the Moon."[73] He makes clear that should a spaceflight vehicle exist, humankind will avail itself of its capabilities: "Who would not be a *High-flyer*, to be Tackt and *Consolidated* in an Engine of such sublime Elevation, and which lifts Men, Monarchs, Members, yea, and whole Nations, up into the Clouds; and performs with such wonderous Art, the long-expected Experiment of the Voyage to the *Moon*?"[74] Here Defoe articulates, writing over half a century after the publications of Godwin and Wilkins, how voyaging to the Moon had now become "long-expected" and how such a voyage would represent new opportunities for individuals as well as for nations.

By the end of the first quarter of the eighteenth century, other new ideas and a renewed optimism related to spaceflight began to emerge. Murtagh McDermot's *A Trip to the Moon*, first published in 1727 in Dublin and reprinted in London in 1728, provided a detailed description of an important new technique: the first gun launch into space, a method that would later be made famous by Jules Verne.[75] Almost nothing is known about McDermot, with the name presumed to be a pseudonym, but the story itself and its publishing history point strongly to the author being Irish. In the story, McDermot arrives at the Moon by virtue of a whirlwind that sent him into the heavens from the highest mountain on Tenerife in the Canaries. On the Moon, he makes friends, experiences its stimulating coffeehouse life, and enjoys travel on a lunar submarine before acquiring a death sentence and resolving to return to the Earth. He describes his plan to do so to his closest lunar acquaintance: "We already know, said I, the Height of the Moon's *Atmosphere*, and know how Gun-Powder will raise a Ball of any Weight to any Height: Now I design to place myself in the Middle of ten wooden Vessels, placed one within another, with the Outermost strongly hooped with Iron, to prevent its breaking. This I will place over 7000 Barrels of Powder, which I know will raise me to the Top of the *Atmosphere*."[76] His lunar friend objects that both McDermot and the vessel would be set on fire by this method, but McDermot had already given this some thought:

> Between the Vessels I will pour a Quantity of Water, which will prevent the burning any but the Outermost, by boring Holes in the Sides of all the Innermost, excepting that in which I stand, thro' which the Water may freely pass to quench the Outer, which, I believe, will be fir'd; that which is in the Middle shall have no Holes in it, that so the Water may not come near me: the upper Part of my Tubs I will cover with two Plates of Tin, between which I will place a Quantity of wet Sand to prevent the lowermost Plate from setting the wood on Fire. This Cover shall be fastned by me on the Inside, so as that I can throw it off when I think that I am got out of the Smoak. I'll supply the Want of fresh Air by some Bottles of animal Spirits; and will prevent the Vessels from touching each other, by placing Pieces of Lead between the Sides and Bottoms, which upon the Approach of the Fire, will melt and fall off before they can do any Harm. I believe that my Motion will be pretty swift; but I am used to travel Post, and by this Time can bear to be hurried. I provide my self with Wings, to prevent my too rapid Fall to the Earth; for I believe, were I not thus provided, I should not escape so well as I did in my Fall to the Moon.[77]

Here, in an unassuming, almost totally forgotten work, is one of the eighteenth century's most detailed descriptions of a vehicle for spaceflight, one that would not be surpassed in its similarities to real spacecraft for almost a hundred years. The description contains several advancements in spaceflight conceptualization, including the first detailed description of a gun launch to space, consideration of what today we would call *thermal protection systems*, recognition of the need to preprovision air to breathe, and recognition that Earth's stronger gravity provides additional challenges different from landing on the Moon.

McDermot also gets credit for the first description of landing site selection options upon reentry to Earth: "I flew very pleasantly, till I chanced to come to the *Southern* Parts of *Africk*, where I had like to be knock'd down with the Top of an high Mountain, as the Earth was revolving; but I timely perceiv'd it, and got over the *Ethiopick* Ocean. In that I durst not alight, but flying more to the *South*, I alighted on a pleasant Country, situate in *Terra Australis incognita*, where in a delightful Grove I fell asleep, being much tired with my flying, and almost roasted with the Sun."[78] Fascinatingly, McDermot nods to the fact that this method may not be entirely practicable, but nonetheless expresses hope that better methods will be found in the future: "The Design, though rendered a little feasible, might not succeed, and many unforeseen Accidents might cause it to miscarry. As to that, said I, I must rely upon Providence, which I hope will send me with News to my Countrymen, whose Ingenuity may, in Time, find out a more easy Method of maintaining Correspondence with other Planetary Inhabitants."[79] Although it is impossible to know whether McDermot genuinely believed in the possibility of spaceflight, particularly when knowing so little about author's identity, he nonetheless treated the subject with some seriousness and expressed, like Wilkins and Russen, genuine hope for advances that might one day enable the realization of a real-world spaceflight narrative.

A few decades after *A Trip to the Moon*, another presumed pseudonymous author, Ralph Morris, added some new conceptual advances and a couple of remarkable historical coincidences to the narrative development of spaceflight. Published first in 1751, *The Life and Adventures of John Daniel* contains a voyage to the Moon in a vehicle called the *Eagle*—which would also be the name of the lunar lander for the Apollo 11 mission some 218 years later. The book also includes one of the most charmingly descriptive title pages in science fiction history (figure 1.3). Although the epic book covers a wide variety of topics and adventures other than space

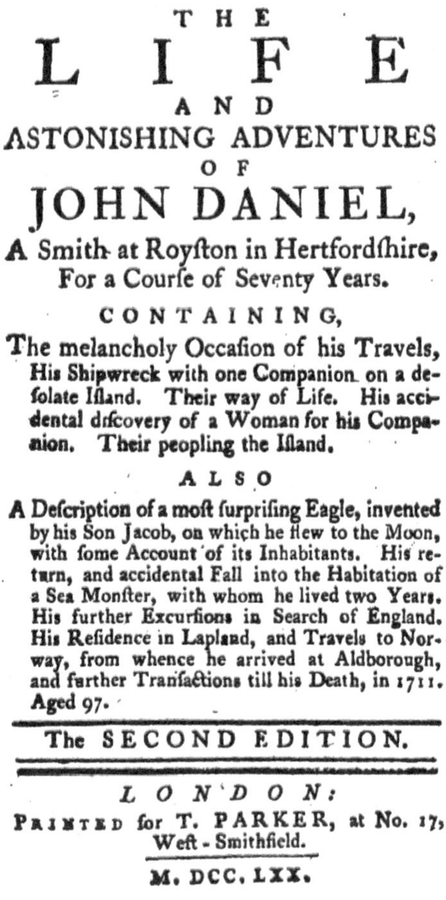

THE
LIFE
AND
ASTONISHING ADVENTURES
OF
JOHN DANIEL,
A Smith at Royſton in Hertfordſhire,
For a Courſe of Seventy Years.
CONTAINING,
The melancholy Occaſion of his Travels,
His Shipwreck with one Companion on a deſolate Iſland. Their way of Life. His accidental diſcovery of a Woman for his Companion. Their peopling the Iſland.
ALSO
A Deſcription of a moſt ſurpriſing Eagle, invented by his Son Jacob, on which he flew to the Moon, with ſome Account of its Inhabitants. His return, and accidental Fall into the Habitation of a Sea Monſter, with whom he lived two Years. His further Excurſions in Search of England. His Reſidence in Lapland, and Travels to Norway, from whence he arrived at Aldborough, and farther Tranſactions till his Death, in 1711. Aged 97.

The SECOND EDITION.

LONDON:
PRINTED for T. PARKER, at No. 17, Weſt-Smithfield.
M. DCC. LXX.

1.3 Title page of *The Life and Adventures of John Daniel. Source:* Ralph Morris, *The Life and Adventures of John Daniel* (London, 1770), title page.

travel, the opening sentence nonetheless implies the cosmic scope of the narrative, describing it as "an account of some of the most surprising adventures that ever befel [sic] any one man in the universe."

John Daniel is son of a blacksmith who joins a sailing ship after being forced to leave England due to the undesired amorous advances of his stepmother. He ends up marooned on a tropical island with a woman—Ruth Comin—who is initially disguised as a man in order to serve as a sailor, and with whom he proceeds to build an ever-growing family on the island. It is one of Ruth and John's sons—Jacob—who begins to develop a new vehicle for flight. Ruth Comin and John Daniel therefore seem to be the first known names of a spacecraft inventor's parents in spaceflight fiction, a minor milestone to be sure, but one which nonetheless highlights the emerging intergenerational nature of spaceflight development narratives.

Jacob is also the first fictional spacecraft inventor whose narrative we follow from childhood. He is noted as someone "whose head was ever after something new" and who has an immense curiosity about his father's homeland of England.[80] John tells Jacob that they came from a land two hundred days away by boat, which leads Jacob to conclude he will never be able to sail there; Jacob thereafter secretly begins using his smithing skills to construct a flying machine. John, meanwhile, spends most of his time aiding his children, and their growing families, while establishing "new settlements" on the island and managing their growth and interrelations. Jacob continues to collect various materials—including wooden poles, wax, and fabric—from the island and builds his ship—his *Eagle*.[81] Jacob's *Eagle* is a platform with secured wooden ribbing attached to waxed fabric wings, all connected to a handle pump in the middle of the platform that, when pushed and pulled, creates the equivalent of the "play of wings in birds."[82]

Jacob and John then fly in the machine, away from the island and through the night—aided fortuitously by the antitiredness and antithirst properties of a special leaf that they chew—and by accident end up flying to the Moon, although they don't realize this at first. They disembark from the *Eagle* and explore by the light of the "moon," which is in fact the Earth, until eventually the Sun rises and they find a wonderful new world: "the most romantick country I had ever beheld," full of beautiful mountains, plains, lakes, and forests, though also "very much broken, with prodigious risings, and as hideous depths."[83] They explore the Moon and encounter its inhabitants before departing—remarking on their floating without weight in the process—and return to the Earth for further adventures, first with a humanoid sea monster, then with inhabitants of Lapland, before returning to England on a whaling ship.[84]

On their return to the Earth, the text begins to get philosophical. In their encounter with the humanoid sea monster, it is revealed to John and Jacob that they have been to the Moon. The sea monster's remarks are suffused with wonderment that such a feat could be achieved by simple humans:

He lifted up his hands and eyes; Great Father, says he, what have I heard? And hath all this been done by a worm! A creature seemingly pinned down to the earth, whose natural bulk, and tendency to the center, should render such attempts a violence to nature? But who can discern the methods of thy providence? And am I, poor miserable creature, the first of they innumerable beings upon this earth, that am to be informed by a special messanger of our neighbour world, which I have so often contemplated, and have been amazed at? Am I, I say, the first, to whom the truth of that new world hath been discovered?[85]

The sea monster even concludes his enthusiastic monologue on the wonders of cislunar travel with a version of the classic phrasing later made famous by *Star Trek*: "Then, says he, you have been where never man was before. You have been in the moon."[86]

Perhaps the most remarkable element of *The Life and Adventures of John Daniel*, however, is its explicit recognition of a religious motivation for space exploration, one that was first hinted at in the writing of John Wilkins: the motivation to go on a space pilgrimage. When John Daniels finally returns to his old home in England, he is over ninety. Jacob had died along the way from a whaling accident, and John makes friends with a local parson with whom he shares knowledge of his son's *Eagle*. The parson is much enamored with the idea of a flying machine, and so John reconstructs the vehicle to show him its capabilities. The parson is duly impressed and explains to John that were he younger, he would assuredly use the *Eagle* to travel into space himself. His motivation, however, is not for commerce or conquest or curiosity, but for spiritual communion with God and the cosmos:

There is not so sublime a notion of the divine Being to be obtained by any other means, as arises from the contemplation of his works, than what a fountain of joy and delight, praise and gratitude, must spring up from the actual survey of them; I mean the remoter parts of them, those as yet, viewable but through a cloud, and only guessed at from imperfect hints and surmises. I would not only have gone to the Moon, our neighbour planet, but to Venus, nay Mars, Jupiter and even Saturn himself, had my years lasted, should have been visited by me. O! I had there seen the order, regularity, and nice disposition of Jupiter's satellites, and discovered the use of Saturn's ring; I had observed the form, the designs, the exercise, and faculties of the several specific of inhabitants; united my voice to theirs, in prasing our great Creator and Preserver; and whether ever

I had returned or not, what had it mattered, so that my soul had been filled with the sense of those mighty works of creation, and of the omnipotent Agent that performed it all.[87]

Here then, with the first explicit description of the desire to go on an interplanetary spiritual pilgrimage in history, we begin to see the broader contours of the many idiosyncratic motivations that motivate people to travel into space.

Although the focus here has primarily been on English writers, throughout the eighteenth century, French writers continued to write boldly about interplanetary travel as well, with one of the most important works coming from one of the future fathers of the French Revolution. In 1752, Voltaire published *Micromégas*, which tells the story of a giant interplanetary natural philosopher who is around thirty-nine kilometers tall. When he is forced to leave his home star system of Sirius at the age of 450 years old, due to poor treatment by local religious zealots, he travels to Saturn, where he befriends a "dwarf," who is only around two kilometers tall, before traveling on to Earth.[88] The work is one of the most expansive investigations of the implications of both the scale of the cosmos and the concept of the plurality of worlds, not to mention the value of scientific inquiry and true humility in the face of the vastness of the universe. Bernard Le Bovier de Fontenelle's *Entretiens sur la pluralité des mondes* (Conversations on the plurality of worlds) in 1686 had popularized the idea of other inhabited planets among French intellectuals, and Voltaire's *Micromégas* both satirized its cautious presentation and expanded on its ideas. Whereas Fontenelle had presented the idea of the plurality of worlds as part of a polite conversation in an earthly garden, Voltaire presents the plurality of worlds as a fact—with truly alien intelligences engaging in conversation as they travel between planets, while pitying the humans they observe on Earth living within a vast universe that they can barely comprehend. Voltaire includes no description of a vehicle for space travel, as both Micromégas and his Saturnian companion are simply able to innately voyage between planets with their massive bodies, and all the space travel is done by nonhuman inhabitants. Nonetheless, the *grand philosophe* Voltaire, in addition to all his other literary and philosophical laurels, can lay claim to having written both the first narrative of interstellar travel and the first multiplanet adventure.

Building from Voltaire's tradition, Madam Marie-Anne de Roumier-Robert in 1766 wrote *Voyages de Milord Céton dans les sept planètes, ou Le Nouveau Mentor*, which is the earliest multiplanet adventure that features a human doing the traveling. Roumier-Robert was from an aristocratic family, and her father knew Bernard Le Bovier de Fontenelle, who she had met on multiple occasions; she correspondingly focuses much of her narrative on social and philosophical observations related to the similarities and differences between the inhabitants of different planets.[89] Although her travel into space is initially made on the wings of Zachiel—a "genius," "genie," or "djinn," depending on translation—she also deserves note for being perhaps the first to suggest traveling through space by what would later become known as *wormholes* when she notes that this is how the genii travel from one planet to another: "'I can see,' Zachiel added, 'that nothing ought to keep us any longer on this world [the Moon]; thus, we can now go on to the planet Mercury. In order to go there more comfortably, however, I shall take you into a vortex; they are the vehicles of which we shall make use for all the voyages that we shall be continually obligated to make in all the possible worlds. . . .' We followed the genius, although slightly alarmed by the aspect of the vortices, which might be compared to chaos."[90] Later, in a further conceptual leap, Roumier-Robert has her protagonists switch from interplanetary travel via vortices to interplanetary travel via the fundamental constituents of the universe: "Judging that there was nothing more to retain us in the globe of the Sun, Zachiel proposed that we take to the vortices again in order to reach the planet Jupiter—which is, as is well known, one of the largest and most distant from our world—or, to have us transported there by atoms. . . . The genius put us on a group of hooked atoms which were aggregated as if linked together. Those atoms, which are not as frightening as vortices, conveyed us quite comfortably to the planet Jupiter."[91] Although the concept of an interplanetary magic carpet made of atoms may not seem to represent major conceptual progress in space travel, it was in some ways a nontrivial innovation, particularly in demonstrating new consideration of how the basic compositional nature of matter might aid in space travel—a line of inquiry that ultimately leads to the *Star Trek* transporter or the *Stargate* stargate, two other methods that arguably function via "hooked atoms." In a similarly creative foray, in 1775,

Louis-Guillaume de La Folie, in his *Le Philosophe Sans Prétention*, described the first spaceflight machine powered by electricity, albeit as the product of the greatest minds of the planet Mercury rather than of Earth.[92] Even for all the conceptual advances that had been made since the days of Godwin and Wilkins, the possibility of space travel still seemed so remote in the eighteenth century that for those thinking about credible narratives for spaceflight, it was easier to imagine that spaceflight would be developed by genies and advanced alien civilizations than by humans.

The invention in 1783 of the first actual vehicle of human flight—the hot-air ballon—by the Montgolfier brothers, Joseph-Michel and Jacques-Étienne, provides a fitting endpoint for this first phase of our survey of the development of the spaceflight narrative. The subsequent enthusiasm for *aeronautics*—a term first applied to ballooning—resulted in a rapid adoption of this new technology of actual flight as the leading literary device for stories about spaceflight. The most immediate adoption was by the pseudonymous Vivenair, who in 1784 published the satirical *A Journey lately performed through the Air, in an Aerostatic Globe . . . To the newly discovered Planet, Georgium Sidus*. Georgium Sidus was the first new planet discovered since ancient times and given its name—*Georgium Sidus* is latin for *Georgian star*—in 1781 by its discoverer, the English astronomer William Herschel, to honor King George III of England. The German astronomer Johann Elert Bode soon proposed the name that would become its official scientific one—Uranus—although British references continued to use Georgium Sidus until the mid-1800s. That Vivenair would publish a story about a trip to a planet just discovered three years earlier in a vehicle of flight invented only the year before speaks to the growing sense of invention and excitement that the last quarter of the eighteenth century held for those interested in flight and the cosmos. Vivenair also highlights the glory-seeking motivations that began to be depicted as increasingly common as spaceflight began to be depicted as less of an accident and more of an explicit objective for literary protagonists: "Urged on, or rather inspired, by the wonderful experiments of late made and performed by those superlative geniuses, Messrs. Mongolfier, Charles, and others . . . animated also, as it must be confessed, by that selfishness, the usual attendant upon glory; I resolved upon constructing in the most secret manner, an Aerostatic Globe; not only with all the improvements of the great philosophers above named, but those also suggested by my own mind, naturally fertile and inventive."[93] The "improvements" noted by the narrator included a component for steering the balloon that allowed him to navigate toward Uranus as he rose, finding on his arrival a society remarkably similar to England's and therefore ripe for the standard skewering satirical comparisons.

Balloon-borne space travel would remain the norm for the rest of the century. *A Voyage to the Moon, Strongly Recommend to All Lovers of Real Freedom*, published by the pseudonymous Aratus in 1793, is another classic example that makes plain the continuing tie to utopian sentiment for lunar voyages amid the emergence of true aeronauts. As you shall see in the next chapter, it was also the concept of balloon-borne voyages to the Moon that would cross the Atlantic and inspire the first American narrative of spaceflight. Spaceflight conceptual innovations would continue in Britain and France through the early years of the nineteenth century, though still predominantly in the form of largely satirical narratives. One notable innovation was a first spaceflight by hydrogen balloon, described by "Mr. Nicholas Lunatic, F.R.S." in *A Voyage to the Moon* in 1808, while Edward Francesca Burney described another early version of a space gun made up of four large cannons in a pioneering work in 1815—*Q.Q. Esq's Journey to the Moon*—that also included a space capsule made more aerodynamic by the use of a partly unfurled umbrella.[94]

There were notably, and surprisingly, few writers outside of the British Isles and France that contributed new spaceflight narratives during this early period. Hans Jacob Christophel von Grimmelshausen provided the first German narrative on space travel in 1659, *Die Fliegende Wandersmann nach den Mond*, although it was simply a translation of Godwin's tale, and Pier Jacopo Martello described a spaceship that flies via "magnetic" energy generated by a pair of amber globes in his 1707 epic poem *Degli Occhi di Gesù Gli* (The eyes of Jesus).[95] Eberhard Kindermann, a Germanic astronomer who believed he had discovered a moon of Mars in 1744 and who would later become the court astrologer to the prince-elector of Saxony and king of Poland, provided a particularly notable contribution as he published a remarkable story in 1745 about five German sailors traveling in a balloon-borne airship around Mars—the first balloon-borne space travel story, and the first ever spaceflight narrative in the vicinity of Mars.[96] Interestingly, Kindermann's sailors do not land on Mars but instead land on its supposed moon, which in the future may be seen as another remarkably prescient coincidence if the first actual human mission to Mars lands on either of its moons—Phobos or Deimos—as some modern space mission planners have proposed.

Kindermann's work also included the first explicit recognition that travelers on their way to Mars would watch the Sun grow smaller and that it would become correspondingly colder around Mars as well.[97] The frontispiece to his pamphlet *Die Geschwinde Reise*—or, to give it it's full title in English, *The Speedy Journey On the Air-Ship to the Upper World, Which Lately Five Persons Have Undertaken to Determine Whether it is True that the Planet Mars Appeared on the Tenth of July of This Year with a Satellite or Moon for the First Time for as Long as the World Has Existed? Reported to the Curious Lower World for Its Amusement and Edification and to Affirm the Veracity of This Matter by the Well-Known Fama*—also includes the first graphical depiction of an interplanetary voyage, complete with illustrations of Mars and its supposed moon, a diagram of its orbit, a balloon-borne spaceship with oars, and terrestrial commentators (figure 1.4).

This remarkable German tale aside, non-Anglo-Franco spaceflight narratives are relatively few and far between in the seventeenth and eighteenth centuries, and of significantly less impact on either their contemporaries or on the broader intellectual history of spaceflight when compared to the many works produced in the British Isles and France over the same period. Why this should have been the case is unclear. One potential explanation might be that as both countries were Atlantic Ocean communities with colonies already developing in the New World, both British and French writers during this time might have been particularly prone to thinking about traveling to "new worlds." However, this does not help explain why similar narratives do not seem as evident in Portuguese or Spanish sources. It is also at least worth noting that the British Isles and France have long had many more cultural similarities, and even greater unity, than is often appreciated. French was the mother tongue of the kings of England from William the Conqueror to Henry IV in the early 1400s, and French continued on in regular usage among the English elite for some time after, with law French still used in the courts at the time of Godwin and Wilkins and for years afterward; it was only outlawed in exclusive favor of English in

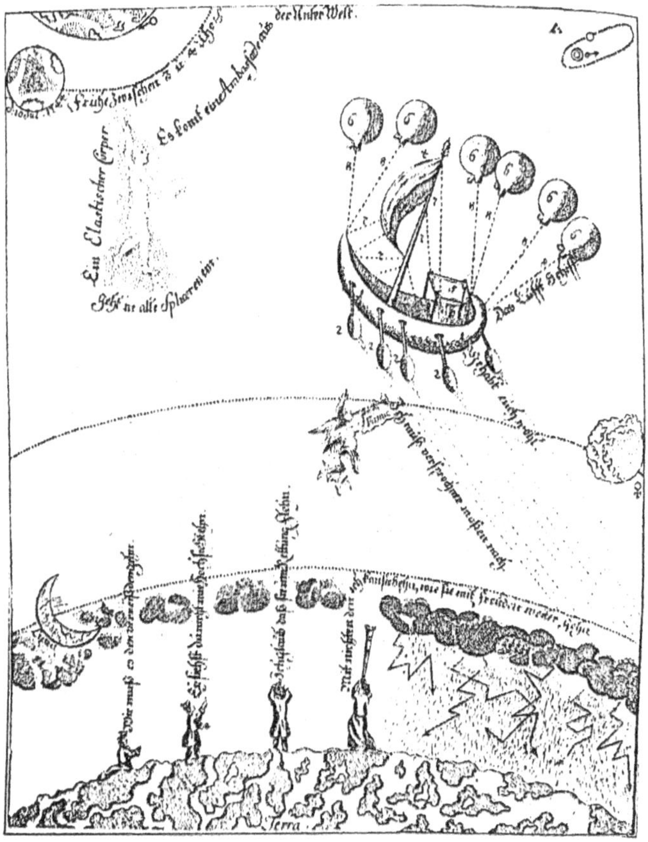

1.4 The voyage to the moon of Mars in *Die Geschwinde Reise*. *Source:* Eberhard Kindermann, *Die Geschwinde Reise* (Berlin, 1774), frontispiece illustration.

the courts in 1730. Regardless of the reasons why, which could certainly afford to be studied further, it is fascinating that this particular corner of the planet, across two languages, generated such a vibrant flowering of thought regarding the potential for the human exploration of the Moon and the solar system.

As you have seen in these earliest examples of the spaceflight narrative, it was private individuals, individualistic invention, and personal interests—as well as very often accidental adventures—that dominated the motivational landscape in these earliest stories, with David Russen's narrative of a massive Moon-reaching spring in the Himalayas being the notable exception. Human journeys into space began as personal journeys, expanding to include a few friends and companions at times, and predominantly were imagined as endeavors that could be undertaken—and, by implication, financed—by individuals, giving the space travel enterprise a distinctly individualistic flavor at its origin. We also see the roots of space expansionist thought at the very origins of the conceptual development of spaceflight. Kepler wrote to Galileo about future space settlers heading to the Moon and Jupiter; Wilkins wrote about the future potential for Earth-Moon commerce; and multiple writers across European spaceflight literature in the seventeenth and eighteenth centuries wrote about planetary and interplanetary empires and the potential for new colonies on new worlds. Rather than considering space expansionist thought as something that begins in the early twentieth century, we should understand that the seeds for space expansionism began to be planted at the very moment that the telescope allowed humans to begin visually exploring and discovering new worlds in the heavens.

At a more fundamental level, however, it was these stories that laid the foundation for the conceptual development of spaceflight technology. Many of these narratives—those by Godwin, Wilkins, and Cyrano de Bergerac in particular—would be directly referenced and cited as inspiration, and as intellectual foils, by the authors of spaceflight narratives in the nineteenth century, who would pick up the spaceflight theme and carry it further than ever before. The intellectual history of spaceflight starts here, then, in the age of candlelight and the Age of Enlightenment. With the dawn of the nineteenth century, however, and the dramatic technological advances of the Industrial Revolution, the realm of the possible was rapidly expanding, and the concept of human spaceflight would once again become a matter for serious technical consideration—this time with many of the most significant advances occurring across the Atlantic Ocean.

2

THUS OPENING AN INTERPLANETARY COMMERCE

Every year the world seems to me to be growing smaller and smaller:
Steamliners across the oceans, Atlantic and Pacific and Mediterranean;
Railroads across America and Europe;
Telegraph from San Francisco to San Francisco all the way around.
If we do not hitch onto the moon and quarry our granite there
it won't be the fault of the Yankees.
—Cleveland Abbe, *Letter from Russia*, 1866[1]

As astronomical observatories became part of the national landscape in the nineteenth century, the increasing enthusiasm for astronomy and the rush of new scientific knowledge led several Americans—buoyed by the rapid growth of their nation, its technologies, and its industries—to imagine an extension of American exploration and expansion out into the solar system. Cleveland Abbe, a former director of the Cincinnati Observatory, assistant to Benjamin Gould, and the first American to study astronomy at the famous Pulkovo Observatory in Russia, indicated as much with his statement, perhaps made only half in jest, that he expected Americans to eventually be quarrying granite from the Moon. By the mid-nineteenth century, spaceflight seemed like a genuine possibility in America, and it would be there that the concept of spaceflight would undergo some of its most important transformations.

Early nineteenth-century American spaceflight literature developed economic notions of spaceflight that explicitly emphasized the personal motivations of individuals and the importance of obtaining capital to achieve it. American intellectuals, inspired by their nation's economic and technological development, resumed the serious contemplation of spaceflight that had begun in seventeenth-century England and France and then started developing new concepts that drew on improved scientific knowledge and practical considerations. During the first half of the nineteenth century, these concepts continued to appeal to the tradition of the brilliant lone inventor, with the individual depicted as the source of all the necessary resources and motivations for spaceflight. Nonetheless, the early nineteenth-century American conceptual development of spaceflight was more technically sophisticated than is generally appreciated: It laid the foundation for Jules Verne's *From the Earth to the Moon*, and it planted the seeds for a distinctly American perspective on a human future in space—one led by individuals, ecstatic and utopian desires, and private capital.

In the beginning, however, the first American narrative of spaceflight showed significant continuity with the narratives in Europe at the time. The earliest spaceflight story written in America yet found is one that features deeply religious French protagonists traveling to the Moon in an air balloon. Entitled *An Account of Count D'Artois and His Friend's Passage to the Moon* and published in 1785 by a twenty-one-year-old Daniel Moore from Connecticut, it was written after American independence but before the creation of the US Constitution.[2] The book is a predominantly philosophical and religious narrative about humanity in the cosmos and the potential to return to a prelapsarian state, as seen through the lens of enthusiastically devout New England Protestantism.[3] The book takes the form of a letter, written by the French Count D'Artois and sent back to Earth from his new abode in the Moon. The letter describes the story of how he and his friend, a monsieur J. Vogrill, adventured in the heavens and accidently flew to the Moon in a seemingly standard air balloon from Paris.

Shortly after the start of their journey, they find themselves twenty miles above the Earth: "Here the most lofty mountains appeared but small rising spots of ground, and almost upon a level with the vallies; and every thing to be seen, afforded us the most pleasing and delightful scene that ever I beheld in the material world."[4] After enjoying this scene, the aeronauts attempt to return to Earth but find themselves unable to do so; instead, they continue on into the blackness of space. They fear for their lives, until Vogrill delivers a speech on the "great goodness" of God that gives them courage, and they accept their unplanned cosmic adventure and the loss of everything they have known of their lives on Earth.[5] On the third day of their voyage in the heavens, supported only by the four gallons of wine that they had brought with them, they discovered a "supposed planet" full of beautiful-smelling trees, with birds, and surrounded by water—and yet somehow existing entirely in the air and supported "on nothing."[6] They are then reminded of scripture, remembering that "He maketh the skies his dwelling place, and rideth upon the wings of the wind," and, along with many other biblical observations, imply that they have passed the original Garden of Eden, from which humanity literally descended to Earth.[7]

Although D'Artois and Vogrill are nominally French, they nonetheless exhibit a distinctly New England Protestant enthusiasm for the spiritual elements of astronomy and sky watching. They share this in common with other prominent early Americans, such as Benjamin Franklin and John Quincy Adams.[8] The sentiments provoked by their journey into the cosmos, of seeing the literal heavens, are of both wonder and horror. They start with cosmic wonder: "O, what a small part of the universe is the lower world! O, how vain a thing is man! For, when called to pass through the skies, and to behold their beauty, how does he hide his head, not being able to endure them? He says, *Come ye mountains, and rocks, hide us from the glory of true excellence.*"[9] In contrast, they later pass by a comet that they think could be a literal hell: "But, should there be a fixed abode, where such lost souls are to languish out a long eternity, I should think this comet to be it which we passed; as it appeared dark and gloomy—horror seemed to spread itself on all its parts."[10] Religious observations and sentiments form a majority of the book, and in many ways *An Account of Count D'Artois and His Friend's Passage to the Moon* is more of an innovative religious tract that uses a space voyage as a method to explore faith and evangelical Christianity than it is a consideration of space travel or a contribution to its conceptual development.

And yet there is much in the book that gives the sense of a real journey, and of a journey to the Moon in particular. On their approach to the Moon, which turns out to be a verdant, paradisical world, they first see the general outlines of human civilization—buildings and people, but also cattle, horses, sheep, and birds. Vogrill then delivers probably the first passable *bon mot* of American space fiction when, on approaching the surface of the Moon, he remarks: "Dear Count, we shall in a few minutes land in a world of which we are wholly unacquainted; but let it not ruffle us."[11] The journey to the Moon, voyaging past literal and metaphorical heaven and hell, takes them a total of seven weeks and six days—a duration for human presence in the cislunar environment that remains unsurpassed in reality.[12]

On the Moon, the count and his friend wait for two hours and then proceed out to the lunar surface. There they deliver a rather extended prayer as the first words on the Moon:

> Immediately after we were out of the Balloon, my friend fell prostrate, and with a loud voice said—"O thou great Original of all worlds, and all beings—that daily observes the same—that daily doth great wonders that are past the conceptions of any created being!—to thee do I now address myself, hoping to be heard in my desires; now, though in a world where we never before had trod. We have been taken, by the hand, from our native world, and placed where we now are attempting to call upon thee;—where there is a numerous people with whom we never have conversed, and of whose feelings we are ignorant;—but to thee are open all their hearts, and who can turn them as water, and cause them to receive us as friends.[13]

What it lacks in the brevity and humanness of Neil Armstrong's famous statement, it perhaps compensates with its generous American expression of hope for good relations with lunar inhabitants.

After this prayer, they return to the balloon and wait until morning before venturing further into the promisingly fertile new lunar landscape. After journeying from the balloon for a while, they come across a yellow brick building and are both delighted and surprised to hear Hebrew emanating from it. They enter the building and find a group of men praying who are initially concerned about the potential ill omen that these new flying visitors represent, but they are greatly relieved when they learn of D'Artois and Vogrill's peaceful and pious interests. The adventurers from Earth soon find the Moon to be a genuine paradise, ultimately concluding that "Eden of old, perhaps, could not boast of greater advantages, of greater beauties."[14]

After spending time on the Moon and finding the lunar inhabitants to be very agreeable, they invite two of them onto their balloon for a short trip to see Mercury, Saturn, and Venus before returning back to the Moon. The two lunar inhabitants are particularly impressed with the balloon spacecraft and the adventures it enables, and they are sufficiently impressed by it that they half wonder whether D'Artois and Vogrill may in fact be immortals: "Here in this place is the greatest piece of true surprizing workmanship of men, that ever our eyes beheld: it is what our fathers were ignorant of, as they have not transmitted to us any thing of the kind: it is without doubt the greatest invention that ever mortals (if so you be) perfected; as the very heavens, by its assistance, may be minutely surveyed, and all their true and glorious beauties—beauties that those that have not seen can have no idea of."[15] Here in the first American spaceflight narrative, the main value acclaimed for the first space vehicle is not its potential to enable voyages of glory or conquest, nor its potential to open up new worlds for material benefit or commerce, but rather simply its ability to enable mortal humans to be able to experience and minutely survey the unimaginable beauty of the cosmos.

After finding such a welcoming new lunar community, D'Artois and Vogrill decide that they will stay in this new paradise on the Moon and send back to Earth an account of their journeys. They close their account with a reminder to their fellow Earthlings to "foresake their sins" and "return to the ways of righteousness" before ending with a poem that provides the first pean to spaceflight inventors and makers in American literature:

O noble Genii, thought exalted high,
That form'd a structure for to range the sky!
May heaven protect such sons of fame,
And earth spread far the great proclaim.[16]

Although earlier European spaceflight narratives frequently discussed religious themes and sentiments, *An Account of Count D'Artois and His Friend's Passage to the Moon* is the first fictional spaceflight journey in which spiritual concerns are the primary focus of the narrative. Religious and faith-based motivations for supporting the American spaceflight enterprise have been identified by scholars in the twentieth and twenty-first centuries, and it is worth noting that these motivations are also observed in the very first American spaceflight narrative ever written.[17]

The next recorded American suggestion of interplanetary travel is much more well known, and it reveals a decidedly more secular and skeptical

American view on the potential of exchange between the Earth and the Moon. In 1809, as the Napoleonic Wars raged in Europe, James Madison replaced Thomas Jefferson as president of the United States, and Robert Fulton filed his first patent on the steamboat, a twenty-six-year-old Washington Irving published the work that would launch his career as one of the most successful and renowned of early American writers: *A History of New York by Diedrich Knickerbocker*.[18] The young Irving—who had met his namesake, George Washington, at the age of six—had been born into a prosperous merchant family. After a two-year social tour of Europe, he had returned to find his chosen path of legal scholarship tedious, and so he turned instead to writing and publishing political and social satire. *A History of New York* was his masterpiece in this vein, lampooning the self-regarding New York society he was surrounded by, as well as many of the dominant American cultural assumptions of the time, including the assumed rightful superiority of Americans of European ancestry and the justice of their territorial expansion and violent campaigns against the continent's Native Americans. To shift the perspective of his European American readership, Irving made the creative leap of describing what it might be like if European Americans were to face foreign invaders who were yet more technologically and militarily advanced than them—foreign invaders from the Moon. Although his principal literary purpose was to satirize contemporary society, in so doing Irving introduces a provocative thought experiment regarding the potential development of means to travel between worlds:

> Let us suppose, then, that the inhabitants of the moon, by astonishing advancement in science, and by profound insight into that lunar philosophy, the mere flickerings of which have of late years dazzled the feeble optics, and addled the shallow brains of the good people of our globe—let us suppose, I say, that the inhabitants of the moon, by these means, had arrived at such a command of their energies, such an enviable state of perfectibility, as to control the elements, and navigate the boundless regions of space. Let us suppose a roving crew of these soaring philosophers, in the course of an aerial voyage of discovery among the stars, should chance to alight upon this outlandish planet.[19]

What is most striking in this passage is that Irving presents the idea of travel between the Moon and the Earth in a reasoned and logical way. There is no technical content in Irving's description, but it is nonetheless implied that the advancement of scientific development could ultimately lead to spaceflight. The process of the advancement of science is stated as a phenomenon that can lead to the control of the elements of the universe to such an extent that travel between the worlds becomes possible once a population's "energies" are focused upon it. Irving also places the power of flight and the first landing on Earth in the hands of "roving philosophers" who are otherwise engaged in a "voyage of discovery among the stars"—providing the earliest American suggestion that the spirit of scientific discovery would play a leading role in motivating voyages into space, exactly 149 years before the flight of Explorer 1.

Writing at a time when candles and oil lamps were still the only forms of artificial lighting available, Irving recognizes that he is engaged in some wild speculation, but he defends it on the grounds of historical perspective:

> I am far from indulging in any sportive vein at present; nor is the supposition I have been making so wild as many may deem it. It has long been a very serious and anxious question with me, and many a time and oft, in the course of my overwhelming cares and contrivances for the welfare and protection of this my native planet, have I lain awake whole nights debating in my mind, whether it were most probable we should first discover and civilize the moon, or the moon discover and civilize our globe. Neither would the prodigy of sailing in the air and cruising among the stars be a whit more astonishing and incomprehensible to us than was the European mystery of navigating floating castles, through the world of waters, to the simple natives.[20]

Although we don't know how seriously or anxiously Irving actually considered the question of lunar invasion, it is clear that he had given the matter some genuine thought with an interest in the practical, and potentially disastrous, consequences of contact with an alien civilization. Although surrounded by satire, he presented the concept of interplanetary travel as plausible, indicating the fertile ground that the seed of spaceflight would find in the minds of the young American nation. Perhaps even more significantly, his explicit connections between the violent European conquest of the Americas and interplanetary colonialization indicate that popular concerns and critiques of space expansionism were beginning to take shape as early as the 1800s.

Four years later, another interplanetary tale would emerge that would provide another heady mixture of the spiritual and utopian themes that would find such resonance throughout nineteenth-century American spaceflight narratives. George Fowler's 1813 work *A Flight to the Moon; or, the Vision of Randalthus* worked with themes of grand cosmic perspective and utopian yearning, as did Moore's *Count D'Artois*, though this time with a firmer appreciation of the science of astronomy.[21] *A Flight to the Moon* is the second American book about a lunar journey, though it is less advanced in its technical conception of spaceflight than the first—with the voyager, Randalthus, being transported to the Moon via the aerial cloud of an angelic figure in a dream.[22] What it lacks in technical consideration, however, it makes up for with an emphatically earnest American attempt by its author to convey his observations of the universe, the Earth, and humanity from a cosmic perspective.

This cosmic perspective, as in the case of the voyage of Count D'Artois, is a deeply spiritual one. Inspired by the notion that the different shades visible on the Moon "represented mountains, forests and oceans," Randalthus exalts the Creator's heavens:

> Ah! How small a portion of the brilliant and boundless universe is it the destiny of feeble and short sighted man to behold! How is he lost in the immensity of creation! How is he even surrounded with darkness through which it is equally beyond the power of the sight and the imagination to penetrate! Yet is not the portion of the universe we behold, the too dazzling sun and the beautiful moon . . . sufficient to direct our thought to the omnipotent Creator, and to inspire feelings above a groveling attachment to this earth.[23]

This reaction to the heavens is more than plain religiosity; it is scientific understanding that has most strengthened Randalthus's fervor. He refers to astronomy as "the true temple of God's worship" and even expresses a pilgrimatic planetary wanderlust when he declares that "I long to break my chains and pour out my effusions on each of the planets that wing their rapid flight in their eternal rounds."[24] When Randalthus arrives cloud-borne to the Moon, he informs its people—which he finds less scientifically advanced than the people of Earth—that "the enlightened part of the species of my native world, tearing themselves from trivial concerns, and conversing only with suns and worlds, have discovered the bulk and distance of [the] planets."[25] The awe-struck reaction of the Lunarians to these statements presumably reflects the wonderment felt by Fowler himself—and that of many other Americans in the early republic—when confronted with the vast reality of the solar system and the universe that had been revealed by recent astronomical discoveries.

Fowler, through Randalthus, also comes down strongly in favor of the plurality of worlds, while admitting to some empirical uncertainty in the matter at the same time: "To affirm that they are inhabited, O men of the moon, would be to affirm, what nothing but an actual visit to them, could realize. . . . The philosopher mourns that a truth, and a glorious truth apparently within his reach, should yet elude his grasp; and sighs in the sublime, but uncertain idea, that he will at some future day, contemplate those worlds in the brightness of truth, and feast his feelings on the wisdom of their divine organization."[26] Even the scale of this insight, though joyfully enrapturing to the Lunarians, is not enough for Fowler: "What we have been considering is only the skirts of the creation; and, in relation to the extent and grandeur of the universe, no more to the enlightened astronomer, than the confused and distant view of a bank or a tree would be to the mariner thirsting to explore the regions of the vast continent to which they belonged."[27] For Fowler, this grand cosmic perspective, as for many later space expansionists, is simultaneously humbling and thrilling on a personal level, and unifying and leveling at an interpersonal one, and thus it provides grounds for the human Randalthus and the Lunarians to embrace as brethren.

However, Fowler's attention is as focused on the failings of human society as it is on the grandeur of the universe. Randalthus and one of his Lunarian companions ascend a high lunar mountain, and there Randalthus points out each of the visible regions and nations of his distant home planet, opining on their respective histories and failings in turn. After being overwhelmed with the apparently disheartening state of affairs on planet Earth, the Lunarian asks Randalthus a question that resonates to this day: "Is it possible that creatures who have unfolded the mysteries of nature and developed the laws by which distant bodies move in their eternal rounds cannot understand the principles of their own nature, or adopt by mutual consent a system for their general and individual benefit?"[28] Even though Randalthus holds an apocalyptic belief in the likelihood of a divine cataclysm to come, there is nonetheless a yearning and hope in *A Flight to the Moon* that a better world and social structure—one that could avert the anticipated apocalypse—might yet be possible. Given the effort Fowler expended in expressing his thoughts and fears on the matter, it seems that he hoped that the consideration of a human journey out into the vastness of space might, in some small way, contribute to the promotion of a more harmonious life on his home planet. Although it is at best only a prelude to the American conceptual development of spaceflight, Fowler's enthusiastic tribute to the cosmic perspective shows that even in the early days of the republic, idealism and hope for a better world were at the core of American space narratives.

The American conceptual development of the spacecraft proper begins with George Tucker, an attorney, former US congressman, and wide-ranging writer who was invited by Thomas Jefferson to become the first professor of moral philosophy at the University of Virginia.[29] In his first years at the university, Tucker wrote a short novel, *A Voyage to the Moon*, published under the pseudonym Joseph Atterley in 1827.[30] The novel contains a description of a spacecraft that represented major conceptual improvements, including descriptions of a technology for the provision of air on the voyage and technical countermeasures against the vacuum and cold of space. Although *A Voyage to the Moon* is fairly well known, it is less well appreciated that it also represents the first American description of the economics of spacecraft construction.

In a number of respects, *A Voyage to the Moon* is the culmination of the early type of lunar voyage narrative that began with Francis Godwin's *The Man in the Moone* and that was elevated briefly to scientific interest by John Wilkins.[31] Even in the early nineteenth century, the seventeenth-century Englishmen's works were still the go-to references for spaceflight, being referred to in the first lines of the *American Quarterly Review* article on *A Voyage to the Moon* written anonymously by Tucker's close friend and colleague Robley Dunglison.[32] Both Godwin and Tucker's stories also share strong narrative parallels, with both featuring highborn first-person narrators who are shipwrecked on a distant shore and, while marooned, discover the secrets of lunar travel. But whereas Godwin's Domingo Gonsales builds his makeshift goose-borne engine for simple flights around his island and only accidentally visits the Moon when the geese migrate there, Tucker's Joseph Atterley learns the secrets of spaceflight from a venerable Indian scientist who designs and then contracts to build their spacecraft for the explicit purpose of exploring the Moon.

The unnamed "venerable Hindoo" is a member of an ancient lineage of Brahmin scientist-priests from Benares, India—today Varanasi—and it is through science, albeit tinged with a healthy dose of mysticism, that Tucker describes the development of the spacecraft as having been realized. In the story, the scientists of Benares are described as the most advanced and secretive in the world, having "discovered the secret of gunpowder, the magnetic attraction, the properties of electricity, long before they were heard of in Europe."[33] It was at this ancient "seat of science" that the Brahmin learned from another learned sage that there was a type of metal that embodied the principle of *repulsion*, a physical property analogous and opposite to the principal of gravitation in the earth.[34] The scientists of Benares conducted "a number of curious experiments" with the metal, learned that it was attracted to the Moon, and proceeded to construct flying carriages of it, allowing them to achieve three landings on the lunar surface.[35] Joseph Atterley therefore is not the engineering hero of the story; he is but a passenger. The technical hero—the first spacecraft engineer in this particular narrative version of the spaceflight universe—is the unnamed Indian scientist who not only completes three lunar voyages with his spacecraft but also, near the end of his life, decides to make one final trip. For this journey, he takes along a recently shipwrecked American, evidently for no other reason than simple generosity and good cheer and that the American seemed particularly keen on the notion.

Although the concept of an antigravity metal is not specifically scientific, as was publicly recognized in the day by Tucker's colleague reviewer Dunglison, it was nonetheless through a process of scientific investigation and experimentation that the spacecraft is described as having been invented. Scientific instruments also fill the spacecraft: "The Brahmin took with him a thermometer, two telescopes, one of which through the top of the machine, and the other through the bottom; a phosphoric lamp, pen, ink, and paper, and some light refreshments sufficient to supply us for some days."[36] Although this outfitting of the spacecraft is remarkably modern, it is the engineering description of the spacecraft itself—for it is a true spacecraft, arguably the first ever described—that has earned *A Voyage to the Moon* its lasting fame:

> The machine in which we proposed to embark, was a copper vessel, that would have been an exact cube of six feet, if the corners and edges had not been rounded off. It had an opening large enough to receive our bodies, which was closed by double sliding panels with quilted cloth between them. When these

were properly adjusted the machine was perfectly air-tight, and strong enough, by means of iron bars running alternately inside and out, to resist the pressure of the atmosphere, when the machine should be exhausted of its air, as we took precaution to prove by the aid of an air-pump. On the top of the copper chest and on the outside, we had as much of the lunar metal (which I shall henceforth call *lunarium*) as we found, by calculation and experiment, would overcome the weight of the machine, as well as its contents, and take us to the moon on the third day. As the air which the machine contained would not be sufficient for our respiration more than about six hours, and the chief part of the space we were to pass through was a mere void, we provided ourselves with a sufficient supply, by condensing it in a small globular vessel.[37]

Here, for the first time in literature, is a fictional spacecraft that approximates the physical spacecraft to come, to such an extent that Atterley's description of awaiting takeoff could almost have been written by an astronaut from the Mercury, Gemini, or Apollo programs: "I found myself shut up in this small chest, and about to enter on a voyage so new, so strange, and beset with such a variety of dangers."[38] Tucker's principal innovation was his airtight copper spacecraft with its fortifying ribbing, making it in effect a pressure vessel—features that he recognized were required to protect the passengers from the vacuum of space. One wonders whether this particular solution—which bears a striking similarity to the copper boilers common in the steam engines of the day—came to the politically connected Tucker in part because of the debates over the steam engine patent rights of American inventors Oliver Evans, Robert Fulton, Nathan Read, and Samuel Morey, which had raged in Congress during the previous decade.

The copper pressure vessel was not the only technical solution that Tucker proposed for meeting the challenges of spaceflight. He also anticipated a need for compressed oxygen and had the Brahmin scientist employ "a chemical process for evolving heat" to warm the spacecraft against the cold of space.[39] Indeed, while the majority of the book is devoted to satirical commentary on the imagined lunar society, as with Irving, the spacecraft innovations Tucker proposed were meant to add plausibility to the notion of spaceflight and thereby heighten the effect of this satire. There was evidently some pride in this, with Dunglinson commenting in his review that, apart from the antigravity metal, "the remainder of the machinery, for reaching that luminary, would not be inconsistent with probability or the known laws of physics."[40] Although this may have been a friendly exaggeration by Dunglinson, Tucker had indeed proposed the first seemingly plausible concept, relative to the knowledge of the day, since the eighteenth century. The advancement of American technology and industrial development had once again made it possible to imagine an engineering solution to the problem of spaceflight.

As with any engineering program, there are attendant economic considerations, and Tucker is diligent enough to include them. Once Atterley and his Brahmin companion had agreed to embark on their journey together, the Indian described his program plan: "We have a great deal of preparation to make. Our apparatus requires the best workmanship, and we cannot here command either first-rate articles or materials. . . . I know an ingenious artificer in copper and other metals. . . . But we must give him time."[41] In fact, they had to give him three months and a "moderate compensation" from the gold pieces the Indian scientist had been saving for the occasion.[42] The spacecraft, with its rare metal, is thus a product within the means of an individual, but not one that is entirely trivial in its expense. Amazingly, Tucker even provides the first-ever suggestion of space program cost calculations: In conversation with the people of the Moon, the Indian scientist, endeavoring to allay their concerns of a possible invasion from Earth, "entered into some calculations, to show that the project was also impracticable on account of its expense."[43] Tucker's lunar spacecraft is thus presented as being economically practicable for a resourceful scientist and an individual who, like Atterley, was motivated by "the love of the marvelous, and the wish for a change," but not for a grand project of conquest.[44]

A voyage to the Moon, in this early American conception, is thus a somewhat costly proposition, but one that can result from the right combination of scientific intelligence, personal motivations, and individual resources. The former congressman had outlined a nearly plausible description of a spacecraft and some of its basic economic requirements. But it would take the pen of a tortured poet to delve deeper into the science and psychology of leaving the earth and to produce a work that would both set a new standard for the verisimilitude of interplanetary tales and delve at length, for the first time, into a private obsession with spaceflight.

It may be tempting to dismiss out of hand the notion that Edgar Allan Poe—popularly conceived of today as an almost mythic figure cloaked in ravens and tell-tale hearts—was the "grandfather of realistic spaceflight fiction" and had a significant effect on the conceptual development of the spacecraft.[45] But it is no tall tale. The young orphan who walked the night-lit streets of Baltimore, the first American city with gas streetlamps, had a mind that strove to encompass not only the macabre but the entirety of the cosmos. His greatest work, in his own opinion, was not one of his popular mysteries or phantasmagorias but his prose poem *Eureka*, which originated as his 1848 lecture titled "The Cosmography of the Universe."[46] His tale of a voyage to the Moon, "The Unparalleled Adventure of One Hans Pfaall," first published in 1835, was almost as dear to his heart, claiming that it was the first plausible application of scientific principles to the problem of travel between worlds.[47] Although the boastful Poe oversold his claim, his was indeed the first to remove the veil of satire and focus on the details of the voyage itself. His was also the first, perhaps unsurprisingly, to explore some of the darker aspects of the psychological motivations for spaceflight, and he was the first to situate his spacecraft firmly within the context of mundane economics and engineering. The result was that "Hans Pfaall" became an animating spirit of interplanetary adventure to others—most notably Jules Verne, who, thirty years later, inspired by its visitation upon him, would follow in Poe's gothic wake to the Moon.

As a famed American original, the study of Poe has engendered a voluminous literature, as has "Hans Pfaall."[48] From the perspective of spaceflight history, however, several of its important contributions remain underappreciated. Poe was the first to write a lunar voyage that had the process of spaceflight itself as the principal interest, rather than sociopolitical satire or religious exaltation. In all, Poe's descriptions of the preparations, as well as of the voyage itself, take up the majority of the forty-six-page story, with Pfaall arriving at the Moon only on page forty-three. He also took pains to situate his work within the context of the spaceflight concepts that had come before. In his own "Note on Hans Pfaall," a document as significant as the story itself—as it is the first American essay on the conceptual development of spaceflight—Poe references and reviews his predecessors, from Lucian to Tucker. The rather arrogant Poe was decidedly dismissive of almost everyone—calling Tucker's voyage "deplorably ill conceived" and Cyrano de Bergerac's tales "utterly meaningless"—with the exception of Godwin, whose work he felt was "somewhat ingenious."[49] Poe's opinions and

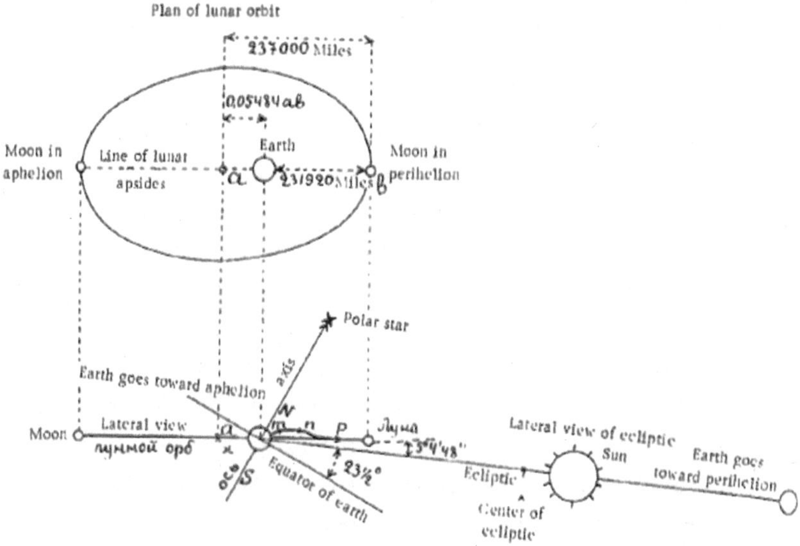

2.1 Scheme of Hans Pfaall's balloon flight to the Moon by Nicolai Rynin in 1928. *Source:* N. A. Rynin, *Dreams, Legends, and Early Fantasies,* vol. 1 of *Interplanetary Flight and Communications* (Israel Program for Scientific Translations, 1970), 94.

proclamations here are less interesting than the fact that there is already a recognition, even in the Baltimore of 1839, of a community of ideas related to the problem of spaceflight that had been developing for generations.

Although he considered himself in conversation with these other spaceflight authors, albeit somewhat belligerently, the principal influences for "Hans Pfaall" were scientific ones. It was Poe's enthusiasm for contemporary astronomy that motivated his interest in the concept of spaceflight, and this had similarly inspired Richard Adam Locke's famous Great Moon Hoax in the same year. Indeed, Poe treated John Herschel's popular *A Treatise on Astronomy*, published in America in 1834, as his textbook, relying on it as his primary source of information about the distance, angles, and orbits required for a voyage the Moon.[50] Although a small step, he was the first to take a specific astronomy text and use it to provide a rough technical description of a lunar mission architecture—one that was sufficiently detailed that it could later be diagrammed by Nikolai Rynin in his pioneering 1928 spaceflight encyclopedia (figure 2.1).[51] Poe would claim that the design in "Hans Pfaall" was "original, inasmuch as regards an attempt at verisimilitude, in the application of scientific principles (so far as the whimsical nature of the subject would permit), to the actual passage between the earth and the moon."[52] To the extent that this was the case, it was due significantly to his application of contemporary astronomical knowledge.

Although the engineering approach presented in "Hans Pfaall" for getting to the Moon—an enclosed balloon-borne carriage with a pneumatic air condenser—has often been dismissed as technically naive, it was actually grounded in Poe's reading of contemporary astronomy. For example, the pneumatic air condenser—which is the core innovation of Poe's spacecraft, and which flows from Hans Pfaall's profession as a bellows maker—is able to function because, as Poe points out, recent astronomical observations had suggested that interplanetary space might have a very thin atmosphere that could potentially be condensed into breathable air with the right technology. For this, Poe points to the zodiacal light and to Benjamin Valz's observation of the decreasing period and diminished coma of Encke's comet when it was near the Sun.[53] Valz proposed that this was the result of a resisting medium, an atmosphere that surrounded the Sun, and Poe's Pfaall simply extended this notion, positing that the zodiacal light was this atmosphere: "It appeared to me evidently in the nature of a rare atmosphere extending from the sun outwards, beyond the orbit of Venus at least, and I believed indefinitely farther. Indeed, this medium I could not suppose confined to the path of the comet's ellipse, or to the immediate neighborhood of the sun. It was easy, on the contrary, to imagine it pervading the entire regions of our planetary system, condensed into what we call atmosphere at the planets themselves."[54]

Other references to the theories and observations of contemporary astronomy abound. When, during his nineteen-day journey to the Moon, Pfaall encounters explosive debris, he presumes it to have come from one of the many volcanoes on the lunar surface, in accord with the volcanic explanation for the moon's pockmarked appearance that was the dominant theory from the eighteenth century to the late 1890s.[55] When justifying his belief in a lunar atmosphere, Pfaall draws on the observations of Johann Hieronymus Schroeter, Johannes Hevelius, and Giovanni Domenico Cassini that suggested its existence.[56] By citing contemporary astronomers and their speculative theories, and thinking about how they would apply to a voyage to the Moon, Poe gave the concept of spaceflight a new basis in science.

In some respects, however, Poe's insights into the psychological and behavioral characteristics that would lead to spaceflight were even more insightful than his clever applications of contemporary astronomy. Although he is still working within the lone-inventor model employed by Godwin and Tucker, the inventor is no longer an aristocratic adventurer, an extraterrestrial, or a mystic scientist, but rather a simple engineer—Hans Pfaall, an underemployed and debt-ridden Dutch bellows maker in Rotterdam. Even more significantly, this inventor begins to approach the level of single-minded obsession and dedication of effort and resources that would possess many of the early pioneers of spaceflight. Like many of them, Pfaall has an almost epiphanic spaceflight inspiration after reading a book—in this case, *Speculative Astronomy* by Professor Encke:

> There were some particular passages which affected my imagination in an extraordinary manner. The longer I meditated upon these, the more intense grew the interest which had been excited within me. The limited nature of my education in general, and more especially my ignorance on subjects connected with natural philosophy, so far from rendering me diffident of my own ability to comprehend what I had read, or inducing me to mistrust the many vague notions which had arisen in consequence, merely served as a farther stimulus to imagination; and I was vain enough, or perhaps reasonable enough, to doubt whether those crude ideas which, arising in ill-regulated minds, have all the appearance, may not often in effect possess all the force, the reality, and other inherent properties of instinct or intuition.[57]

After a sleepless night of thinking, the next morning Pfaall uses the last of his money to buy some books on mechanics and practical astronomy and "devoted every spare moment to their perusal."[58] When he decides to build his space vehicle, Pfaall sells all his remaining property and "borrowed, in small sums, under various pretences, and without giving any attention . . . to [his] future means of repayment, no inconsiderable quantity of ready money."[59] Although, like most of his predecessors, Poe conceived of a spacecraft as something within the technical capability of a single individual, he also foresaw that the challenges of building a spacecraft would require resources beyond what the average person would be able to raise under ordinary circumstances, and he was the first to depict a spacecraft inventor as an obsessive individual endeavoring to secure funds at any cost in order to realize his vision.

Importantly, the process of spacecraft construction was an integral part of Poe's narrative and was presented as a relatively standard engineering problem. Poe described in exhaustive and minute detail the articles and instruments that Pfaall needed in order to build and equip his spacecraft. The result was what can arguably be considered the first spacecraft components list (figure 2.2). Poe also situated the construction of the spacecraft firmly within the mundane context of real engineering, with Pfaall concerning himself with material costs and adjusting his design accordingly. He selects cambric muslin on account of it "answering all the purposes of silk itself, quite as strong and a good deal less expensive," while rejecting a "membrane of a certain animal" for the balloon as he thought it "altogether too expensive and was not sure, upon the whole, whether cambric muslin with a coating of gum caoutchouc was not equally as good."[60] This attention to practical engineering detail would be the greatest legacy of "Hans Pfaall." When Jules Verne—who considered Poe "le createur du roman merveilleux-scientifique" (the creator of the marvelous-scientific novel) and his personal literary hero—described his massive engineering project in *From the Earth to the Moon*, his emphasis on the practical, almost pedantic aspects of the project is the logical conclusion of Poe's movement toward a real-world context for spaceflight.[61] In 1835, however, it would still be thirty years before the spirit of Poe would reach across the Atlantic to inspire Verne's foundational work. In the meantime, the idea of an interplanetary balloon would have an even more detailed and innovative realization in the mind of a professor of chemistry in the medical department of the University of Louisiana, a man who would become the first trained American scientist to write on the subject of spaceflight and who would deliver, in antebellum New Orleans, the first public lecture on the topic in recorded history.[62]

Dr. John Leonard Riddell, born in Leyden, Massachusetts, in 1807, was one of the greatest minds of mid-nineteenth-century American science. Long forgotten in the historical literature, his reputation and achievements have only recently begun to be revisited.[63] Of his incredible career—which encompassed heights in science, invention, and politics—only a general overview can be given here. With his training in botany and geology from the Rensselaer School in New York, Riddell started his career as a professional itinerant lecturer giving public lectures on scientific subjects in Ogdensburg, New York; Kingston and Brockville, Upper Canada, now Ontario; and Pittsburgh, Pennsylvania, from 1830 to 1832, before moving on to Ohio

Hans Pfaall's Spacecraft Components List
Sheets of Cambric muslin, very fine, 12 yards each
A lot of varnish of caoutchouc
1 large and deep basket of wicker (made to order)
Twine
1 hoop
Cords
1 quadrant
1 compass
1 telescope
1 common barometer with some important modifications
1 thermometer
1 electrometer
1 magnetic needle
1 seconds watch
2 astronomical instruments 'not so generally known'
1 stick of sealing wax
1 bell
1 speaking trumpet
1 globe of glass, exhausted of air, closed with a stopper
5 iron-bound casks (50 gallons each)
1 larger iron-bound cask
6 tinned ware tubes, 3-inch in diameter, 10-ft long
A quantity of a particular metallic substance, 'which I shall not name'
12 demijohns of a common acid (makes 'special light gas' in combination with the 'particular metallic substance.')
1 of M. Grimm's improvements on the apparatus for condensation of the atmospheric air
175 lbs of ballast
Numerous instruments and materials for experiment in the upper regions of the upper atmosphere
Several other articles necessary in the construction and equipment of a balloon of extraordinary dimensions
A copious supply of water
A large quantity of provisions, such as pemmican, 'in which much nutriment is contained in comparatively little bulk'

2.2 Hans Pfaall's spacecraft components list. *Source:* Edgar Allan Poe, "The Unparalleled Adventure of One Hans Pfaall," in *The Works of the Late Edgar Allan Poe* (New York, 1850), 1–51.

and lecturing in Marietta, Worthington, and Cincinnati, all while conducting some of the first botanical expeditions in the state and earning his MD at Cincinnati College. The active young scientist then moved to New Orleans in 1836, where he soon established himself as a respected public figure in the second-largest port city in America. He performed the first botanical and geological survey of the Edwards Plateau in Texas, invented the practical binocular compound microscope, and conducted pioneering bacteriological investigations of cholera and yellow fever in the 1850s. He also became the melter and refiner of the New Orleans Mint, the New Orleans postmaster, a member of the Louisiana River Control Commission, and was even elected and sworn in as the rump governor of Louisiana in a citizens' protest of the state's then-military governorship during the tumultuous final years of the Civil War.[64] In addition to all of this, on April, 30, 1847, Riddell also took to the podium of the People's Lyceum of New Orleans and delivered a discussion of a "new mode of aerial navigation, embracing a narrative of a voyage to the moon." With that act, he became the first in history to deliver a public lecture on spaceflight, and arguably the first scientist to seriously consider the possibility of building a spacecraft since John Wilkins in the seventeenth century. From an economic perspective, Riddell was also the first American to introduce that perennial dream of modern space expansionists: interplanetary trade.

Riddell's spaceflight considerations were more than a passing interest. As early as 1831, while lecturing in Ogdensburg, he had penned a stately note in his journal: "Send an expedition to the moon. A wonder that in the present age of enterprise no expedition should have been projected to the moon which is so near to us."[65] To the perceptive young scientist, who had grown up amid the nation's early adoption of steam power and the railroad, the advancement of science and industrial enterprise seemed to make an expedition to the Moon an obvious idea. Riddell would later give voice to this youthful sentiment in the published version of his lecture, in which the hero of his story, Orrin Lindsay, reveals his motivations: "In early life I had allowed my fancy to go wild on the subject of aerial navigation; I envied the birds of the air in the possession of their talent of flying, and longed to be able to imitate them. I fondly hoped and believed, that human science and skill would at length attain this desideratum."[66] Already, before even the electromagnetic telegraph had achieved widespread usage, the Moon was beginning to seem genuinely within reach to at least one far-seeing American.

Because Riddell's spaceflight narrative is such a significant milestone in the American conceptual development of spaceflight, and because it is so poorly known in the space community, it is worth covering at length. Riddell's lecture at the People's Lyceum was so successful that his friends and admirers encouraged him to publish it, with the result being the booklet entitled *Orrin Lindsay's Plan of Aerial Navigation, with a Narrative of His Explorations in the Higher Regions of the Atmosphere, and His Wonderful Voyage Around the Moon.* As with his lecture, the format of the book is not satire, but rather an earnest description of the technical aspects of spaceflight using, with only one important exception, the known science of the day. Riddell distances himself from the idea somewhat by framing the narrative as a series of correspondences with a Mr. Orrin Lindsay, a man who was "shy and eccentric in his habits" but nonetheless "manifested great capacity, zeal, and talents for the acquisition of truth."[67] To lend the tale of his correspondence greater credibility, Riddell also claims—in a verisimilitudinous touch that would have made Poe envious—that Lindsay was an acquaintance and a student of his public lectures when he was in Cincinnati.

The core of Orrin Lindsay's spaceflight concept, and Riddell's singular departure from known science, is the idea of creating a "gravity screen," based on the notion that gravity is "the result of a direct influence which matter exercises upon matter" and that it might thus be possible to conceive of "compounding and preparing a substance of such a molecular structure, as to be to this influence impervious."[68] Although Riddell, as the lecturer, is careful to distance himself from this statement and claim to grand invention by "Lindsay," he nonetheless provides a discussion and footnotes on the nature of matter, the laws of motion, and gravitation to lend the idea credence. He also defends the idea by analogy to general scientific advancement, stating that "if fifty years ago, some eccentric genius in advance of his age, had asserted that the fearful spirit of the thunder storm might be enslaved by man, and made to transmit human thought, in a moment's time, across the breadth of a continent; who would not have said, inconceivable and absolutely impossible! Yet we now see all this actually accomplished by the magnetic Telegraph."[69] Before presenting Lindsay's account of his invention and voyage to the Moon, Riddell thus invites his audience to suspend for a moment their disbelief and to consider what might be possible.

With Lindsay's "Brief Account of Some Novel Experiments Upon Gravitation. And Also a Narrative of Two Voyages into Empty Space"—which comprises most of the text—Riddell is able to create an aura of authenticity for his spacecraft. He does this by providing a more realistic depiction of the process of invention and construction than was achieved by any of his literary predecessors in the genre, no doubt drawing on his own experience as a practicing scientist and inventor. Lindsay begins his account by noting that it had been six years since he first reflected on the "vast reservoir of momentum" that must exist in the "inscrutable principle" of gravitation.[70] He had thus experimented for six fruitless years on the subject, almost giving up many times before he finally achieved success. He found that "well prepared steel," after being amalgamated with mercury and then strongly magnetized, became a type of screen, impervious to the influence of gravitation. He then conducts experiments, first with a one-foot-cube box with sheets of the gravity-screening metal on all sides in order to levitate the object, and with holes placed in the metal to allow the box to move in the outward direction of the opening. He then experimented with a "young bull terrier dog" in a larger square box attached to the ground with twine, carefully measuring the velocity of the dog's ascent until the twine broke and the dog and box ascended out of sight—a sacrifice to the science of flight, as would be the Soviet space dog Laika 110 years later.[71]

With the experimental aspects completed, Riddell then extends and elaborates on the lone-inventor model of spacecraft production and adds a technical assistant for the construction phase of the project. Intending to develop a vehicle with what he assumes will be significant commercial applications, Orrin Lindsay removes himself to a remote location—a small house at the bottom of the "Devil's big Punch-bowl" outside of Natchez, Mississippi—where he sets up his "special assortment of artisans tools, as well as considerable chemical and philosophical apparatus."[72] He does not move alone, however, bringing with him a single employee: a technical assistant, Mr. Abner Josslin, "a skillful philosophical instrument maker of Cincinnati."[73] Lindsay and Josslin are also assisted by Dr. T. B. Osserman, a paleontologist who accidently discovers their workshop while searching for mastodon bones, and "Messers Wilkinson and Harmason, proprietors of the Natchez Foundry," who construct "isolated pieces of machinery" for the team.[74] Here then, with the return of a scientific mind to the problem, the American conceptual development of spaceflight begins to approach the type of activity that would be seen with Robert Goddard's first liquid fuel rocketry experiments some eighty years later: a skilled scientist with skilled technical assistants establishing a purpose-built workshop at a remote site in order to construct their spacecraft.

After four months, Lindsay declares that their "experiments, trials and careful mathematical researches" have been a success and that the vehicle—which he calls a *magnetic balloon*—is ready for flight. The first description of a spacecraft prototype from an American scientist is thus provided:

> Its external form was globular, and ten feet in diameter. It consisted of a strong frame of wood work, covered with poplar boards, upon which, externally, were secured plates of the magnetic amalgam. There was one considerable opening, in the manner of a door, into the interior, externally also covered with the amalgam; and twelve circular spy-holes, or windows, symmetrically and equidistantly placed over the whole surface of the sphere, each one six inches in diameter, with twelve plates of the magnetical amalgam, to close the openings at pleasure. Within, the furniture consisted of a small low table, a chair, a box of philosophical instruments, &c., &c.[75]

After conducting rope-bound trials of the vehicle to the height of the Devil's Punch-bowl, Lindsay enters the vehicle, closes the windows apart from those perpendicular to the Earth, has Josslin loosen the rope, and the vehicle departs into the air on its first crewed test flight. On this inaugural voyage, the scientist Riddell is interested principally in describing scientific phenomena, such as the effect of zero gravity on the candle he brought along for lighting, the real-world behavior of which he correctly predicts: "The flame of the taper became globular and less luminous; the heated air from it, seemingly not knowing which way to rise, diffused itself slowly in every direction."[76] Even more time is spent on his observations of the temperature and atmospheric density as the vehicle rises to five miles in altitude. At this altitude, his breathing becomes quick and laborious, telling him that he will need to make alterations to his vehicle if he is to brave the expanse of space. He then returns to his camp to make the necessary preparations.

In preparation for the second voyage, during which Lindsay intends to visit the Moon, Riddell has Lindsay perform, for the first time thus far known, the types of calculations required to solve the problems of environmental control and life support in space. Lindsay's first task is to line the vehicle with furs and blankets, to mitigate against the cold of space, and with an airtight lining of sheet lead for the hull and airtight plate glass for the twelve openings, to counter the vacuum. He then calculates how much oxygen will be required for him and Josslin. Based on his assumptions that a healthy man consumes 23.75 cubic feet of oxygen in twenty-four hours, he constructs two strong copper reservoirs with a capacity of 24 cubic feet, and he compresses the oxygen by force pump to 1/20 of its normal volume to provide for two men for ten days.[77] Lindsay notes that this creates an interior pressure for the reservoirs of three hundred psi, and he adds brass stopcocks for releasing the oxygen. He also notes the danger of the exhalation of carbon dioxide at the rate of two pounds per day. To absorb the forty pounds of carbon dioxide, Lindsay calculates that they'll need sixty-eight pounds of "slacked lime," or calcium hydroxide—a compound still used to absorb carbon dioxide in diving equipment today. To absorb the moisture from the respiration of their skin and lungs, Lindsay adds a carefully positioned box of "unslacked lime"—that is, quicklime, or calcium oxide. To round out their provisions, Lindsay then throws in water, food, and, for good measure, "a few bottles of wine and a bottle of brandy."[78] No previous description of the provisions required for a trip to the Moon had ever reached this level of detail. Riddell, with his training in chemistry, had designed the first crude environmental control and life-support system for a spacecraft.

Riddell's description of the journey to the Moon represents a similarly significant milestone, including as it does the first consideration of an atmospheric aerocapture approach to planetary landing and a prescient description of a bleak and desolate Moon. There is even an allusion to the idea of using a "gravitational slingshot" for increasing the spacecraft's velocity, with Lindsay and Josslin leveraging the gravitational attraction of Earth to pick up "tangential velocity" to travel the 240,000 miles to the Moon in a just few days rather than the forty-four that would be required using the gravitational pull of the Moon alone.[79] As the pair travel the expanse between the worlds, Riddell begins to wax poetic, while staying true to his scientific background: "The temperature of this region, if it may be said to possess temperature, is far colder than we had means of measuring. We were here presented with no azure sky, no diffuse daylight as on the earth. The sky was intensely black, and multitudes of stars ever visible whether the sun shone or not, glittered with a brilliancy never seen through the earth's atmosphere, like diamonds upon a ground of black velvet."[80] In planning his trajectory, Lindsay accounts for the Moon's relative motion and targets its limb, where he hopes to put into practice his plan for atmospheric aerocapture by "gliding along horizontally through the moon's atmosphere, so as to get rid of our immense velocity without harm, preparatory to landing on the moon's surface."[81] He discovers, however, that there is no meaningful lunar atmosphere present, and the pair instead end up in an elliptical two-hour orbit around the Moon—another first—at a mean distance of one hundred miles and reaching only twenty to thirty miles above the surface at their closest approach.[82]

From his orbital position, Lindsay observes that on the Moon there is only an "exceedingly thin and light atmosphere, no more than 1/3 of a mile from her average surface," and that the atmosphere was probably "more rarified than in the nearest approach to a vacuum which we can produce in our air pumps."[83] In a footnote, Riddell performed a number of detailed calculations, based on observational data from the astronomer Elias Loomis, to show why this would be the case. Whereas Poe had sought a coating of scientific verisimilitude, Riddell sought nothing less than the application of genuine science. The result was a depiction of a Moon that was no longer an Edenic playground for social satire and potential encounters with alien forms of life and society, but rather a stark and dead world: "Not a drop of water was any where visible. There were no indications of animal or vegetable life; but all seemed a cold, dry, dreary, wild and barren waste."[84] In antebellum New Orleans, barely a decade after the Great Moon Hoax and its imagined lunar inhabitants had transfixed the attention of New York and the nation, the stark reality of the Moon was coming into view.

This marks a significant milestone in the evolution of spaceflight narratives: the moment when lunar narratives cease to assume that the Moon will be not only habitable but also already inhabited by extraterrestrial life. Today, it can be hard to imagine what an extraordinarily enticing object of exploration and potential commerce the Moon would have seemed when it would have been broadly assumed to be an extant home of extraterrestrial civilizations and a potential ready abode for enterprising humans. Indeed, for the first two centuries of humanity's consideration of voyages to the Moon, there are essentially no narratives that present descriptions of the Moon as we know it today: thoroughly uninhabited and sufficiently hostile to human life that significant technological advancements and major capital investments are required to maintain life there for even a matter of months. From Kepler and Wilkins to Tucker and Poe, all had assumed that there would be intelligent life on the Moon and presented the Moon as habitable in their narratives. From 1610 to 1847, human interest in lunar exploration was synonymous with interest in encountering the types of extraterrestrial life and society that were imagined to be present there. Although a number of stories would still imagine the Moon to be inhabited and habitable for decades to come, by the end of the century, most writers would follow Riddell and adopt the scientific consensus of the time in their narratives and, in so doing, move their imagined extraterrestrial civilizations to farther abodes—specifically, to Mars and Venus, which would become the new objects of interplanetary interest and hope in the spaceflight narratives to come. Riddell's story thus represents a significant turning point, with his unprecedentedly accurate scientific description of the Moon and the two centuries of hoped-for encounters with lunar civilizations that his lecture began to leave behind.

Although Riddell's description of the Moon is the most striking to us today, his description of Lindsay's return to Earth was no less insightful and carefully considered. After 4.5 orbits of the Moon, Lindsay and Josslin, having realized that the "apparent want of an adequate atmosphere, induced us to give up all idea of attempting to land," opened the spy-holes in their vehicle to be drawn back to Earth.[85] Riddell evidently understood something of the challenging nature of atmospheric reentry, and he provided a description that at least hinted at the dangers inherent in this phase:

> So exceedingly great was our velocity that I feared for the consequences of the impact. We had a clear moonlight, and were moving horizontally west. We felt a slight shock, and perceived a flash of light as we passed through the aerial billow; the friction of the balloon upon the air being sufficient to produce that effect. In a few moments our meteoric light ceased, and again after some twenty minutes had elapsed, the light re-appeared accompanied with another slight shock; and thus for perhaps eight or ten times at decreasing intervals, the shocks and the luminosity returned, both becoming at each successive return less and less perceptible. These singular phenomena, I presume were produced by the balloon's rebounding from the aerial billows, just as a cannon ball will ricochet along the surface of the water.[86]

Impressively, Riddell here seemingly anticipates a particular type of atmospheric reentry called a *skip entry*, where the spacecraft skips across the atmosphere—like a flat stone skipping across the surface of water—which he accurately describes as being possible in the thinner, upper atmosphere. As Lindsay and Josslin proceed, fearing coming down too quickly in the heavier atmosphere, Lindsay opens one of the spy-holes to lessen the rate of the descent. When they land back on Earth, outside of Natchez, Riddell provides a description that could have been used for the first cosmonauts and astronauts: After having been "hermetically sealed up for the space of six days, twenty hours and thirty minutes," they were relieved to "once more set foot in the open air upon *terra firma*, thankful that we had escaped in safety, the perils of the great deep of ethereal space, in which we had had the temerity to sojourn."[87]

Thus ends Lindsay's account, with the lecturer Riddell adding that "whatever may be thought of its authenticity, [it] certainly embraces and connects together, many authentic and important facts in Natural Science."

But the innovative Riddell is not quite done with his revelations, for there is an important postscript added to the tale. Riddell reports that he later received another letter from Lindsay, which outlines the financial concerns of the young inventor, as well as his plan to use his invention to open up a new frontier of commerce:

> I find that the pecuniary outlays with which I have prosecuted these experiments, have made such considerable inroads upon my moderate fortune, that I can no longer say I possess a competency. I therefore solicit your advice, as to the most proper and ready mode of realizing something from my invention,—being fully resolved, as soon as I can command adequate means, to construct a balloon upon a more extended scale, and make a voyage to the planet Mars. The telescope has revealed to us that like our earth, Mars possesses a considerable atmosphere, and that his surface undergoes the changes of seasons. In short, analogy would lead us to infer that he may be peopled by intelligent beings, perhaps not inferior to ourselves. I have a strong desire to go and see; and if a refined and rational people are to be found there, who knows but they may be ten thousand years in advance of us in arts and sciences? And who can foretell the advantages which would accrue to our race, from thus opening an interplanetary commerce?[88]

With his scientific considerations of spaceflight concluded, the refiner and melter of the New Orleans Federal Mint added one final provocative question: If all this proceeding thought and speculation has any grounding in fact, might not the prospect of space travel someday prove to be a real economic boon? With the term *manifest destiny* having been coined only two years earlier, the eyes of Americans were already being turned to Mars with thoughts of further expansion and wealth.

More eyes might have been turned still if Riddell had published his tract more widely or engaged in correspondence with others on the subject. Instead, however, the booklet remained largely confined to his friends and admirers in New Orleans until it began to be rediscovered in the public libraries of the early twentieth century. One of those who rediscovered Riddell was Noel Deisch, a patent draughtsman in Washington, DC, and an early member of the American Interplanetary Society, who looked to Riddell for both inspiration and technical knowledge. Deisch cited Riddell in his pioneering 1926 manuscript *The Navigation of Interplanetary Space: A Study of the Question (Problem) of Flights to Other Worlds*, as well as in a technical paper he wrote on oxygen provision for a space voyage that he submitted to the first International Astronautics Award competition in 1928, for which he received an honorable mention.[89] That Riddell's text would serve as a technical citation in an early astronautics paper eighty years after he wrote it speaks to how far ahead of his time he was. How the American conceptual development of spaceflight might have proceeded if his tract had been more widely known earlier is an interesting, though unanswerable, question. Regardless, Riddell is another genuine intergenerational link in the technical development of astronautics.

Minds across America would continue to turn to the heavens, even without the stimulation of Riddell. As the excitement of the California Gold Rush was drawing the avaricious and the adventurous to the wild West Coast of the continent, and ambitious telegraph engineers were beginning to formulate their plans to meet the great challenge and opportunity of a transatlantic cable, the first concept of a rocket-powered spacecraft was quietly emerging from the mind of a young New York writer.

The juvenile novel *Gulliver Joi* is noteworthy both for its almost complete obscurity and because it contains the first unambiguous description of fully rocket-propelled spaceflight.[90] The forgotten American who penned this prescient work was Elbert Perce, a New Yorker raised in Buffalo; he wrote *Gulliver Joi*, his first book, at age twenty.[91] The novel, first published in 1851, was a youthful imitation of *Gulliver's Travels* that achieved significant commercial success and had "enjoyed quite wide popularity," according to *The New York Times*.[92] Perce himself was a wide-ranging polymath: He translated a number of popular Swedish books; compiled an eight-volume collection of statistics on "all the notable battles ever fought" entitled *The Battle Roll*; and invented his own patented *magnetic terrestrial globe*—an educational product that bore his name and for which he was best known, which placed magnets inside a globe in order to allow small metal figures to attach to its surface.[93] His obituary in *The New York Times* also suggests that he was a spaceflight pioneer with a rather pleasant and unassuming character: "Exceedingly modest and unobtrusive in his deportment, the simplicity and sincerity of his character made all those who knew him his firmly attached friends."[94] Perhaps his quiet personality provides some explanation for the curious fact that the earliest conceptual description of interplanetary rocket-propelled spaceflight was stated plainly at the beginning of an American children's book and then left unpromoted by either its author or others for almost a century and a half.

The tone of *Gulliver Joi* is appropriately rollicking and adventurous, and Perce wastes no time in setting up the modern invention that enables his interplanetary Gulliver's travels. The young, adventurous Gulliver Joi longs to see the world but is shipwrecked on his first voyage; luckily, he is saved by an old man who almost instantly uses a powerful telescope to show Gulliver a new, beautiful planet and promises him part of his great fortune if he will follow his instructions and travel there. Gulliver is told to focus his efforts on studying astronomy, which he does for six months while the old man is engaged in the heating and tempering of a strange molten substance. The old man's efforts end in success when an explosion produces a small heap of red powder, at which point he produces a large hollow cylinder made of an "exceedingly light substance, yet as hard as iron itself" and declares that it is "the vehicle that is to convey him to the new planet."[95] The old man then proceeds to set up his launch platform and vehicle:

> Having all his utensils at hand, he affixed the cylinder in a frame, attached to which were several very strong steel springs. These springs he drew down to their utmost tension, and placed them at the sharpened end, the other extremity of the cylinder pointing upwards. . . . Then taking the red powder that had been formed by the explosion of the crucible, he placed it in a strong, square, steel box, from which, at the end, there issued a small, but very strong tube, and fastened the box securely in the pointed end of the cylinder. He thus explained the principles and the manner of working the machine: "The red powder which I have just enclosed in the steel box, and placed within the cylinder, is a pyrotechnic compound of

great power. As long as a steady heat can be obtained enough to keep it in fusion, so long a steady blast of exceedingly powerful flame will issue from the tube of the steel box, which tube you perceive extends through the aperture at the pointed extremity of the cylinder. For the purpose of keeping up this heat, I have, after great labor, and much research, invented the malleable flame, which I shall place securely under the steel box that contains the powder. This small stop-cock is for the purpose of diminishing the blast whenever you wish."[96]

Although written in the language of juvenile literature, there is a remarkable sophistication in not only the design of the spacecraft overall, but also in this first description of the operation of a rocket engine for spaceflight. Navigation is aided by a double telescope and achieved by virtue of a strong weather vane–like magnet that is drawn to the large quantity of iron in the distant planet. The old man helps Gulliver into the cylinder, the interior of which is lined with fur and has enough room for him to lie comfortably and access the telescope, vane, and stop key. The old man then places the globular "malleable flame" under the box, at which point "instantly a stream of fire issued from it, striking the rock with great violence."[97] After giving Gulliver a manuscript of instruction and saying a blessing over him, the young adventurer recounts how the old man pulled the trigger that confined the steel springs, "and propelled by their force, and that of the flame, I shot up into the air, the long broad flame of fire streaming behind me like a blaze of a comet."[98] While en route to the new planet, he switches the orientation of the cylinder so that the flame is pointing at his destination for a time before he shuts it off, by which means he "was enabled to descend towards the planet with ease."[99] There, plainly, in the opening of a 1850s-era children's book, is the uncanny first description of an interplanetary rocket flight.

Although the rest of the book is a series of almost miscellaneous fantasy adventures in the satirical style of Swift's original, Perce's idea of interplanetary travel was the critical hook of the story. *The United States Magazine and Democratic Review,* in its notice of new books, boiled down the essence of *Gulliver Joi* to a tale of "supposed voyages, by means of modern invention, to the planets."[100] Perce himself returned to the theme at the close of his narrative, with his hero deciding to emigrate to the new world with his entire family and their farmhands, thus necessitating a twelve-person spacecraft—the largest yet constructed in American literature.[101] The old man who had launched Gulliver on his initial journey also ends the story with plans to run a cargo and passenger service between Earth and the new planet twice weekly, with Gulliver hoping that it will open "a direct line of communication between the two spheres."[102] Although surrounded on all sides by juvenile fantasy and satire, the spaceflight concepts of Elbert Perce were startlingly original for 1851 and serve as a reminder, as with the original *Gulliver's Travels*, of the sometimes unlikely guise of innovative new ideas. With interplanetary migration and interplanetary commerce also mentioned by Perce, *Gulliver Joi* also reveals the growing frequency of explicitly space expansionist ideas in early nineteenth-century American spaceflight fiction.

Ultimately, however, *Gulliver Joi* was a simple adventure book aimed at young readers. Although it begins with an impressively accurate description of a rocket-propelled spacecraft, by the end it includes giants that breathe hydrogen and golden dragons with table manners. While some mid-nineteenth-century Americans—perhaps including the modest but brilliant Perce—had begun to appreciate that a journey to the Moon and the planets was beginning to border on the possible, in the broader popular culture it remained in the realm of fantasy. Perce's incidental description of a rocket-propelled spacecraft, prescient though it was, did nothing to change this perception. What was missing was not the idea that a rocket could travel into space; this could be supplied easily enough, as *Gulliver Joi* shows. What was still missing was a popular narrative that firmly grounded spaceflight as a potential program of known science, engineering, and industry and that laid out the fullness of how such an endeavor might be realized at the scale of effort that would be required.

What then can we conclude from this first phase of American spaceflight thought? For one, wonder and religious awe in the cosmos were strong sources of early inspiration and motivation. In the American context, Christianity had a particularly dominant role, as exemplified in the works of Moore and Fowler, but it is interesting that even in the early nineteenth century, the Vedic tradition and Hinduism make a prominent appearance in Tucker's narrative as well. Poe introduced darker, more personal, and introspective motivations for spaceflight, and Riddell and Perce clearly articulated commercial motivations. For Tucker, Poe, and Riddell, however, the primary motivations described in their spaceflight narratives were all idiosyncratic, individualistic, and personal. Even in the early nineteenth century, then, there is no single motivation or rationale for spaceflight, but rather a multiplicity of them—as the National Academies would recognize some two hundred years later.[103]

Many of the elements of space expansionist thought expressed in this period had been present in earlier European writings—such as the ideas of interplanetary commerce and interplanetary migration—but there were some new ideas as well. The enthusiastic Christian utopian projections on the Moon by Moore and Fowler come across as distinctly American in comparison to the more secular, though often still religiously minded, earlier treatments in Europe. It was also an American writer that first provided a spaceflight narrative with an implicit critique of space travel, with Washington Irving employing a spaceflight thought experiment of a lunar invasion of Earth to encourage his fellow Americans to consider the implications and consequences of their treatment of the indigenous peoples of the Americas from a different perspective. By and large, however, we see in American spaceflight narratives of the first half of the nineteenth century a broad continuity with types of space expansionist thought that had been seen earlier in Europe.

What perhaps most distinguishes the early American spaceflight narratives from what had come before, and from contemporary works elsewhere on Earth, was the attention to technical detail that began to build as the nineteenth century progressed. Tucker described the first pressure-vessel spacecraft, Poe infused unprecedented levels of technical consideration into his story, and Riddell provided a scientific and deeply considered public speech on the possibility of spaceflight. There was significant enough elaboration of technical details in early nineteenth-century American spaceflight fiction that we can see the beginnings of recognizably modern spacecraft. Even the juvenile fiction of Elbert Perce contained remarkably prescient technical details in the use of a rocket not only for launch but for deceleration at another planetary destination as well.

Nonetheless, the most significant spaceflight narrative of the century would not come from America. Fourteen years after Perce's rocket, it would be a man across the Atlantic in Paris, France, who, significantly inspired by Edgar Allan Poe, would provide a narrative that would offer the first full-scale technical and economic outline of a space program and, in the process, inspire first the pioneers of liquid fuel rocketry and later a global spaceflight movement.

3

FROM THE EARTH TO THE MOON

When I wrote about them as realities these things were already half-discoveries. I simply made fiction out of what became ulterior fact.
—Jules Verne, interview in *The London Daily News*, 1902[1]

Jules Verne was the single most important contributor to the conceptual development of spaceflight in the nineteenth century. He was influenced by what had come before, but he also went significantly further in terms of his technical contributions, cultural impact, and the scale of his influence and impact. The book commonly referred to and published as *From the Earth to the Moon* in English was in fact two books—*De la Terre à la Lune, trajet direct en 97 heures 20 minutes* (From the Earth to the Moon: a direct route in 97 hours, 20 minutes), first published in French in 1865; and a sequel, *Autour de la Lune* (Around the Moon), published in 1869. Verne's combined lunar narrative is considered to be one of the founding documents of spaceflight.[2] These stories foresaw, with unprecedented insight and detail, many of the technical challenges related to traveling to the Moon, as well as the complex socioeconomic dynamics that surround major space missions overall. Its impact was such that in 1902 it served as the primary source of inspiration for the first spaceflight motion picture—*Le Voyage dans le Lune*—over forty-five years after its original publication. Perhaps most importantly, it directly inspired three key pioneers of liquid fuel rocketry: Konstantin Tsiolkovsky, Hermann Oberth, and Robert Goddard. There are few more important works in the entire history of spaceflight, and arguably in the intellectual history of space expansionism, than *From the Earth to the Moon*.

There is no shortage of analysis of the impact of Jules Verne's contributions to the development of spaceflight. There have now been two annotated English-language versions of *From the Earth to the Moon*: Walter James Miller's 1975 annotated translation detailing the myriad technical contributions that Verne made to the concept of spaceflight; and Anastasia Klimchynskaya's 2026 annotated edition, which also uses Miller's translation and which adds modern annotations and critique from a wide array of leading science fiction and spaceflight scholars.[3] Despite these excellent works that document Verne's contributions, the way that Verne presaged both the economic scale of spaceflight projects and the independent organization of spaceflight enthusiasts, is significantly less appreciated, and in many ways just as impressive and important as his technical ideas. As the writer of arguably the most significant spaceflight narrative of the pre-spaceflight era, Verne's life and work is an instructive example of how spaceflight development can be both foreseen and encouraged decades or centuries ahead of its realization.

Jules Verne was born in 1828 into a well-to-do family in Nantes, where his father had a successful legal practice. He grew up in a comfortable lifestyle, the oldest of five children, and displayed an early enthusiasm for wild adventure. At the age of eleven, he conspired with other boys his age to sneak aboard a merchant vessel bound for the East Indies, and even managed to board the ship and get underway on the journey, although his father managed to catch a steam-ferry and retrieve his adventurous son at the next port of call in France.[4] Thus thwarted, he would be sent to boarding school before moving to Paris to study law at the age of eighteen. After graduating, however, instead of pursuing a legal career, he was determined to pursue a different type of adventure: making his living as a writer. In Paris, he had made friends with Alexandre Dumas, the son of the author of *The Count of Monte Cristo* and *The Three Musketeers*, and in time the elder Alexandre Dumas became his first literary mentor. He also became friends with the vibrant explorer Jacques Arago, brother of the noted astronomer Francois Arago, who had published one of the scientific exploration bestsellers of the day, his *Promenade autour de monde,* which described his voyage around the world on the French corvette *Uranie.* Adventure and exploration were thus instilled early into the young Verne's mind.

As early as 1848, Verne encountered the translated writings of one of his great inspirations, Edgar Allan Poe. In particular, his reading of Poe's "The Balloon Hoax" and "The Unparalleled Adventure of One Hans Pfaall" directly stimulated his young imagination and resulted in his first narrative of flight, *Un Voyage en Ballon,* published in 1851, which was a popular science introduction, written for families and children, to the history and emerging techniques of ballooning. Verne's other modest initial literary offerings—which were mainly theater plays—did not lead to early fame or fortune, and the following years were lean ones. By 1852, Verne had lost any flow of paternal funding after declining to take over his father's law practice in Nantes, and he committed himself to working at the unremunerative Lycée Theatre in Paris as a secretary, helping to manage budgets and programming. He soon married, shifted into working as a clerk in a stockbrokerage firm, and had his only son, Michel, while he continued to work at developing his craft of writing. It was in 1862 that Verne, then thirty-four, dramatically changed the trajectory of his literary career when he developed a connection to Pierre-Jules Hetzel—the book and magazine publisher who had brought to the French reading public the works of Balzac, Victor Hugo, and Emile Zola and who would become Verne's mentor, friend, and publisher until Hetzel's death in 1886. The manuscript that Verne had brought to Hetzel in 1862 was *Cinq semaines en ballon (Five Weeks in a Balloon),* which was published in novel form in 1863 and became an immediate bestseller.

Verne did more than just capitalize on the growing enthusiasm for aeronautics with *Five Weeks in a Balloon;* he also became directly involved with its development and promotion. This involvement gave him social and economic insights that would allow him to accurately describe some of the elements of development and promotion that he would later apply to imagining the development process for astronautics as well. In the book, the hero—an Englishman, Dr. Samuel Fergusson—invents a balloon, the *Victoria,* which can utilize different currents of air to steer the balloon and thereby venture toward Africa, the exploration of which was of great public interest in Europe, with British explorers John Speke and

James Grant undergoing their final expedition to search for the source of the Nile in 1862 while Verne was writing the book. This novel method of trying to steer a balloon was of significant interest to Verne's friend Félix Tournachon, more popularly known as Nadar, who was a pioneering portrait photographer for Paris's fashionable elite and who had taken the first aerial photographs in history in 1858. Around the same time as Verne was working on *Five Weeks in a Balloon,* Nadar formed La Société d'Encouragement de la Locomotion Aérienne au Noyen du Plus Lourd que l'Air (the society for the encouragement of aerial locomotion by means of heavier-than-air machines). The organization used Nadar's photography studio as its headquarters, and Verne served as the secretary of the society. Around the time of the publication of *Five Weeks in a Balloon,* the society undertook its most significant public endeavor: the launch of an enormous balloon dubbed *Le Géant,* which was as tall as Notre Dame Cathedral when inflated and which was launched from the Champs de Mars in 1863 to great public fanfare. *Le Géant* was equipped with champagne, elaborate picnic fare, and guests who had paid 1,000 francs for the ride.[5] The balloon landed only a few miles away, but the extensive publicity helped to turn Verne's balloon novel into a true bestseller. Although Verne did not travel in the balloon himself, he eventually would when, in 1877, his own first balloon ascent would be arranged by Eugene Goddard, a well-known balloon aviator friend of Nadar who had been the constructor of *Le Géant.*[6]

Five Weeks in a Balloon was the turning point for Verne's career; it brought him international fame and inaugurated his long relationship with Hetzel. He began to produce new science novels and adventures at a prodigious rate, with a novel about polar exploration, *Les Adventures du Capitaine Hatteras* (*The Adventures of Captain Hatteras*), and one about deep subterranean exploration, *Voyage au Centre de la Terre* (*Journey to the Center of the Earth*), both published in 1864. As he entered this prolific period, he nonetheless found time to create the only literary critique he ever wrote, which was about Edgar Allan Poe. Entitled "Edgar Allan Poe and his Works," it was published in 1864 in the periodical *Musée.* In the article, Verne references many Poe tales, including "Hans Pfaall," which had influenced his balloon narrative and which would also deeply influence *From the Earth to the Moon,* though he noted that Poe could have made his stories more convincing "had he respected a few elementary laws of physics."[7] Verne considered Poe "le createur du roman merveilleux-scientifique" (the creator of the marvelous-scientific novel) and a literary hero, and he correspondingly honored him in *From the Earth to the Moon*, where he depicts his space travelers celebrating both Poe and Hans Pfaall at their headquarters in Baltimore, Poe's hometown, which he almost certainly chose as the homebase for the fictional Gun Club as a tribute.[8]

In *From the Earth to the Moon*, Verne clearly set out to emulate Poe's verisimilitude with regard to spaceflight—and took it many steps further. Verne prided himself on the degree to which he brought scientific knowledge into his novels. Verne's publisher Hetzel would sometimes become frustrated when time-consuming factual research delayed the receipt of finalized manuscripts, but Hetzel was also pleased to promote the scientific erudition that Verne brought to his writing. After *Five Weeks in a Balloon* was so successful, Hetzel and Verne struck a deal for at least one novel to be published per year. Henceforth, Verne's oeuvre was promoted as a continuing series of stories and novels called "Voyages extraordinaires," an overarching brand under which Verne was to publish no less than fifty-four novels between 1863 and 1905, almost all mixing science and adventure-based fiction. In a breathless blurb to introduce the series, Hetzel wrote: "His aim is to sum up all geographical, geological, physical and astronomical knowledge amassed by modern science, and to present the story of our time."[9] Arguably, Verne genuinely lived up to this strident claim when he wrote his fourth extraordinary voyage—following his balloon, polar, and earth's core adventures—which focused on a first human mission of lunar exploration.

From the Earth to the Moon was first published in French in 1865 and then translated into English in 1873. The story centers on a group of American industrialists—in the form of a private society named the Baltimore Gun Club—who, at the close of the Civil War, are seeking to utilize their recently developed heavy armaments–manufacturing capabilities for something new, and who decide with a flourish to create a massive cannon to launch a projectile to the Moon. Led by the club's president, Impey Barbicane, and the club's secretary, J. T. Maston, they raise a vast amount of international funding to develop the cannon and projectile, and they receive assistance from a bold French innovator, Michel Ardan—a thinly veiled version of Verne's friend Nadar—who inspires the Baltimore Gun Club to launch not just a projectile but a capsule that can transport humans around the Moon. *From the Earth to the Moon* ends with the protagonists successfully around the Moon, but their return would await the sequel, *Around the Moon*, in 1870. Indeed, the Apollo astronaut Frank Borman once told Jules Verne's grandson that his wife, having read the first book, was terrified that Borman might never return to Earth—at which point he suggested she read the sequel.[10] Verne's story was sufficiently believable that, when the novel was originally serialized in France, people wrote to Verne asking to be passengers on the flight.[11] The narrative described with unprecedented insight and detail the technical and logistical challenges of traveling to the Moon, and it would both transfer that knowledge and inspire serious reflection on the physics of spaceflight by the earliest pioneers of liquid fuel rockets. Just as interestingly, for arguably the first time ever, it also accurately described the complex and dynamic socioeconomic context that would ultimately surround the first human voyage to the Moon.

The story accurately describes a remarkable number of the technical elements of a lunar voyage. Verne was knowledgeable about contemporary science and technology, but he gave his story even greater plausibility by enlisting the help of his cousin, Henri Garcet, a professor of mathematics and author of a popular astronomy textbook. Garcet provided Verne with technical calculations, including the trajectories and the needed escape velocity, which the book correctly identified as eleven kilometers per second.[12] *From the Earth to the Moon* provided the first detailed technical description of the complexity and scale of effort required to impart such velocity to an object and achieve spaceflight. In so doing, Verne portrayed the challenge of spaceflight as a tractable engineering problem, which would directly influence Tsiolkovsky, Goddard, and Oberth, along with many other early spaceflight pioneers.

As has been frequently noted, the technical details of Verne's spaceflight system were so well-considered that they prefigured several key elements of the Apollo program. In Verne's story, the astronomers of the Harvard College Observatory define the initial parameters of the problem. Their calculations—that an object traveling vertically at a velocity of eleven kilometers per second could reach the orbit of the Moon in ninety-seven hours, thirteen minutes, and twenty seconds—established that "the *theoretical possibility* of the experiment is therefore absolutely demonstrated;

its *success* must depend upon the power of the engine employed."[13] These calculations represented an accurate description of the physics; during the flight of Apollo 11, the velocity en route to the Moon was some 10.8 kilometers per second, with the lunar landing occurring 102 hours, 45 minutes, and 40 seconds after launch.[14] The astronomers also specified that the launch must be within 0–28 degrees of the equator, a constraint that led Verne to place his launch site in central Florida, within a couple hundred kilometers of where the astronauts of Apollo would begin their journey at Cape Canaveral.[15]

Verne's descriptions of the engineering details of the system were comparably precise. His launch system consists of a 275-meter-long gun well, with 181,000 kilograms of explosives to propel the projectile and its passengers on to their trajectory. Although the initial impact of this form of launch would in reality have been instantaneously fatal to any passengers, Verne initially describes the system in the context of an uncrewed projectile. The idea of putting people inside a capsule emerges only halfway through the story, with the introduction of Ardan. As a system for launching uncrewed payloads into orbit, Verne's system was closer to viability than it is often given credit for. Gun-launch systems for achieving orbital velocities—though not original to Verne, as we have seen—continued to receive significant attention, and even significant funding, in the twentieth century, with the gun-launch efforts of Canadian engineer Gerald Bull ultimately reaching space in 1966 with a gun-launched payload attaining a height of 180 kilometers, well above Earth's atmosphere, on a suborbital trajectory.[16] Although there has not yet been a successful orbital gun launch, efforts to realize one have continued into the first quarter of the twenty-first century.[17] The technical specifics of Verne's particular gun-launch system categorically would not have worked, but the general idea in his story of being able to use a gun to launch a payload into orbit is one that engineers continue to believe may be viable and work toward to this day.

Verne's technical insights into the potential characteristics of spacecraft extended significantly beyond his gun-launch concept. Impressively, like Elbert Perce, Verne also famously envisioned the use of rockets in space—specifically as retrorockets to enable maneuvering around the Moon.[18] One of his most prescient and far-seeing engineering insights, however, was the recognition that the physical qualities of aluminum made the metal ideal for use in spacecraft. At the time, aluminum was considered a new precious metal, with a correspondingly high price, and its use for day-to-day engineering was not yet common. Verne's principal character proposed the seemingly fantastical notion of using twenty thousand pounds of it for his spacecraft:

> "Aluminium?" cried his three colleagues in chorus.
>
> "Unquestionably, my friends. This valuable metal possesses the whiteness of silver, the indestructibility of gold, the tenacity of iron, the fusibility of copper, the lightness of glass. It is easily wrought, is very widely distributed, forming the base of most of the rocks, is three times lighter than iron, and seems to have been created for the express purpose of furnishing us with the material for our projectile."[19]

That Verne imagined an aluminum spacecraft using retro rockets to maneuver around the Moon over a hundred years before this in fact occurred is a significant testament to his foresight and to the overall level of conceptual development that occurred in nineteenth-century spaceflight fiction.

In laying out his unprecedentedly detailed lunar voyage, Verne explicitly recognized his debt to earlier spaceflight conceptualizers. In the address where the voyage is first proposed in his novel, Verne makes clear how his adventurers are building on existing intellectual foundations. First, he recognizes the advances in astronomy that had made the consideration of the voyage possible: "'The moon, gentlemen, has been carefully studied . . . her mass, density, and weight; her constitution, motions, distance, as well as her place in the solar system, have all been exactly determined. Selenographic charts have been constructed with a perfection which equals, if it does not even surpass, that of our terrestrial maps. Photography has given us proofs of the incomparable beauty of our satellite; in short, all is known regarding the moon which mathematical science, astronomy, geology, and optics can learn about her.'"[20] Next, his hero recognizes the work of Godwin, de Bergerac, and Poe, who had "'penetrated the secrets of our satellite" through "'experiments which I call purely paper ones.'"[21] The role of industrial development—in particular, in the armaments industry—is also celebrated as having made the "'means of arriving [at the Moon] simple, easy, certain, infallible.'"[22] All the ingredients are seen to be in place, with the only remaining requirement to add the "'practical genius of Americans to establish a communication with the sidereal world.'"[23] Thanks to the intergenerational development of the spaceflight narrative that had preceded him and the many "paper experiments" that had preceded his own, Verne sees himself as being able to put forth a genuinely almost practicable approach to achieving this centuries-old human aspiration.

Verne presents spaceflight not as the result of a single, heroic endeavor but as the culmination of the discoveries of many different individuals and communities: astronomers, science fiction writers, and the men and women who had contributed to the progressive development of technology and industrial capabilities. Shortly before his death in 1905, Verne told a London reporter that he had no gift for prophecy and that his descriptions of technologies prior to their full realization were simply the result of scrupulous research and scientific thinking: "When I wrote about them as realties these things were already half discoveries. I simply made fiction out of what became ulterior fact."[24] The technical and conceptual foundations that were laid prior to 1865 meant that Verne could approach spaceflight as a technology that was already half discovered.

The role of *From the Earth to the Moon* in transferring technical knowledge to the early liquid fuel rocketry pioneers has been widely appreciated for years within both science fiction and space history communities. There has, however, been less consideration given to how the novel may have inspired and foreshadowed the social and economic dynamics of spaceflight. The essence of Verne's story is one of human behavior—of how a group of people conceived of, funded, and executed a project to launch a projectile to the Moon and of how the world reacted to their aspirations and achievements. While Verne's science fiction predecessors had paid some minor attention to the social and economic context of developing a spaceflight system, Verne importantly makes this the central subject of *From the Earth to the Moon*. Of the first book's twenty-eight chapters, the first twenty-six are devoted to describing the conceptual development, logistical preparations, and overall project-management and fund-raising efforts required to achieve the launch.

Verne's realism extended to a recognition of the massive scale required for such a project, including the funding requirements, which he placed at $5,446,675.[25] The 2024 GDP ratio equivalent value of this amount, calculated from 1865 when *From the Earth to the Moon* was first published, is approximately $15.9 billion dollars.[26] The total cost of the Apollo program has been estimated by NASA at roughly $25 billion in 1969 dollars.[27] This is equivalent to approximately $720 billion dollars in 2024 GDP ratio equivalent terms. Although Verne was off by more than an order of magnitude, he was impressively within two orders of magnitude, and he was the first to delineate such a precise financial estimate of the scale of the endeavor. Indeed, estimates would not improve much on Verne until after World War II. French spaceflight theorist Robert Esnault-Pelterie optimistically estimated a cost of $2,000,000 in 1930 (equivalent in 2024 U.S. GDP ratio terms to $636 million), while Philip E. Cleator, founder of the British Interplanetary Society, made his estimate of $100 million in 1938 (equivalent to $33.5 billion in 2024 GDP ratio equivalent terms).[28] Verne was also the first to describe in detail the process of publicly raising funds for a spaceflight project. The development of the gun-launch system in *From the Earth to the Moon* is an entirely private endeavor, initiated by the members of the Baltimore Gun Club—a private association of individuals from the armaments industry. At a special general meeting, the club's president, Impey Barbicane, declares of the Moon that "'it is perhaps reserved for us to become the Columbuses of this unknown world. Only enter into my plans, and second me with all your power, and I will lead you to its conquest, and its name shall be added to those of the thirty-six States which compose this Great Union.'"[29] The response is immediately enthusiastic. To raise funds, Barbicane and the club do not turn to wealthy individuals or to governments; they instead look to "the financial co-operation of all peoples," appealing for funding in a manifesto addressed to "'all persons of good will upon the face of the earth.'"[30] After Barbicane's speech is telegraphed around the world, a public subscription is opened, and $4 million is subscribed in America within three days, followed by a further $1.4 million internationally. Although the dashing French adventurer Michel Ardan subsequently volunteers to travel to the Moon in the projectile, the enthusiastic initial global support is the result of a proposal that is only targeting an uncrewed launch. Verne describes a scenario in which the world is primed for spaceflight and simply awaiting a credible plan. Verne's experience with Nadar and *Le Géant*, and the incredible public response that resulted, almost certainly influenced this view of the popular enthusiasm for wild new endeavors in flight.

Verne was also prescient in his description of the way that private societies and organizations would form to pursue spaceflight ambitions. As Verne puts it: "Now when an American has an idea, he looks for a second American to share it. When they become three, they elect a president and two secretaries. Four, they appoint an archivist, and now they're in business. Five, they call a general meeting, and the club is officially constituted. As so it went in Baltimore. . . . Just one month after it was founded, the Club already numbered 1,833 active members and 30,575 corresponding members."[31] No previous work of spaceflight fiction had imagined spaceflight societies, and time would prove that such organizations would in fact be critical to the early history of spaceflight and remain important mechanisms for the promotion of astronautics, and indeed space expansionism, today. Verne also wryly commented on the morality of this particular society, noting that it had been concluded that each member had, on averaged, killed 2,375 men through their technology and that "such a figure makes it clear that the sole preoccupation of this learned society was the destruction of humanity for philanthropic reasons and the perfection of weapons as instruments of civilization. It was a gathering of Exterminating Angels who were nevertheless the best fellows in the world."[32] It was therefore through the capabilities and enthusiasms of a private society of weapons developers that Verne imagined humanity first embarking on a voyage to the Moon.

The earliest rocket societies in the 1920s and 1930s—such as the American Rocket Society (ARS) and the Verein für Raumschiffahrt (VfR) in Germany—also followed the pattern of the Baltimore Gun Club in initially seeking to employ a public subscription model for their projects. They also shared the confidence of their fictional predecessors in the degree to which the public would share their personal passion for spaceflight and would open their checkbooks to support it. In Verne's narrative, Barbicane repeatedly assures his colleagues not to worry, that money will pour into their treasury and that "'the millions will not be wanting.'"[33] According to Edward Pendray, second president of the American Rocket Society, members held similarly naive beliefs in the society's early days: "We believed generally that a few public meetings and some newspaper declarations were all that would be necessary to bring forth adequate public support for the space-flight program."[34] Given how well-known *From the Earth to the Moon* was, perhaps the ease with which Verne's Moon mission garnered support in the story seeded an overly optimistic spirit in the early rocket pioneers. In retrospect, however, this type of optimism was also essential to the achievement of their goals, given how far away from them they stood.

Verne also described in detail the national sources of financial contributions and the considerations behind them (see table 3.1). The funding motivations he describes come from a mix of imagined high-mindedness and base national stereotypes. He places the US well into the lead as the home base of the project, and he interestingly also assumes a high level of continental solidarity from South America. He ascribes an impressive amount to Russia, citing its "strong interest in science" and tradition of support for astronomical research.[35] The French support the endeavor as a source of grand amusement, while the Turks provide outsized contributions on account of the moon's prominent role in the Islamic religion.[36] The Spanish contribute little because "science is not held in high esteem in that country," while the Swiss are reticent to contribute because of their analysis of the project on a purely commercial basis: "Switzerland could not see any practical value in the experiment. She could not imagine that sending a projectile to the moon would stir up any business there. It did not seem wise to sink any capital in so risky a venture."[37] Affected by a sentiment of competitive pique, the British refuse to contribute anything in view of the American origin of the project.[38] Stereotypes aside, Verne is astute in assessing that various motives would come into play for such an endeavor and that the incentives would be different for different countries. He recognizes clearly that the Gun Club's appeal could not succeed based on a financially pragmatic sales pitch, nor even a single overriding motivation. Competition, interest in science, religious belief, and pure amusement all play a role in Verne's fictional global fundraising for spaceflight. The motivations for spaceflight in Verne's narrative are diverse and multitudinous, just as they would ultimately prove to be for spaceflight in reality. Unlike how the Apollo program was financed, however, all the contributions imagined in *From the Earth to the Moon* were the result of voluntary, individual action in a nongovernmental, nonmarket context. Verne's novel describes a supply-led project promoted by the Gun Club entrepreneurs within a context in which the principal demand factor is the intrinsic interest of individuals around the world in seeing it happen.

Table 3.1 Space expenditure by nation in *From the Earth to the Moon*

Nation	Value of subscriptions
United States	$4,000,000
South America	$300,000
Russia	$273,000
France	$240,000
Prussia	$175,000
Austria	$95,000
Belgium	$95,000
Turkey	$65,000
Holland	$43,000
Italy	$37,000
Denmark	$22,000
Portugal	$21,000
Sweden and Norway	$55,000
North German Confederation	$13,000
Rome	$7,000
Mexico	$300
Switzerland	$50
Spain	$11
England	$0
Total	$5,446,675
2024 US GDP ratio equivalent	$15,900,000,000

Source: Walter James Miller, ed., *The Annotated Jules Verne: From the Earth to the Moon* (Thomas Y. Crowell, 1978), 66–71.

Verne was thus the first seer of spaceflight to predict and describe the massive public interest that would accompany the first flights into space. Although this may seem obvious in retrospect, previous fictional works on voyages to the Moon had portrayed them largely as the result of solitary inventors working in private, with many of the explorers needing to write back letters to Earth to inform people there that the journey had even happened. There was no previous suggestion of interest on a global scale—presumably in large part because most of them were written prior to the development and expansion of telegraphy as a form of global communication. There was also no sense of the anticipatory public interest in spaceflight that Verne would describe as being a key factor in enabling the mission. Verne paints a picture of public enthusiasm strikingly similar to that generated by the space race a century later in the 1960s. The Gun Club project swells patriotism to new heights, with the American citizenry "seized with seleno-mania," Barbicane becoming "one of the leading citizens of the United States, a kind of Washington of Science," and with the nation holding "no other ambition than to take possession of this new continent of the sky, and to plant upon the summit of its highest elevation the star-spangled banner of the United States of America."[39] Public desire to watch the launch results in a tourism boom in Florida, with the imagined hubbub caused by Verne's five million spectators foreshadowing something of the scene that accompanied the one million people that would gather around Cape Canaveral to watch the liftoff of Apollo 11.[40] Verne's description of the social dynamics even extends to commercial attempts to capitalize on the space explorers' publicity, with the famous circus promoter P. T. Barnum offering $1,000,000 to Michel Ardan to tour with his circus—a proposal that bears more than a passing resemblance to *Life* magazine's eventual $500,000 deal with the Mercury 7 astronauts to tell their personal stories.[41] The broader social and economic implications of spaceflight had never before received such detailed attention and elaboration, nor had the fame and rewards of achieving such a feat been so clearly enunciated.

Verne also foresaw some of the elements of project management, industrial partnering, and regional political lobbying that would later accompany large spaceflight projects. This included his assessment that the technical skills and technological assets required for long-range bombardment would likely make for a natural alliance between the armaments industry and those committed to spaceflight. He also identified some of the ways that procurement and contracting might play out for his fictional space venture. Three separate contracts are awarded by the Gun Club to industry and to academia, for the three principle components: one to the Cold Spring Iron Company for the cannon; one to Breadwill and Company for the projectile; and one to the Cambridge Observatory for the large, mountain-top telescope to follow the projectile in orbit.[42] The members of the Gun Club remain in charge of integration and testing and are savvy enough to establish firm, fixed-price contracts with specific delivery dates and a $100 penalty for every day late.[43] The economic value of being the integration and launch site for the mission is also appreciated by state interests in Florida and Texas—the two states with appropriate potential launch sites below the twenty-eighth parallel—who engage in the type of state-based lobbying that would become a familiar part of NASA's future: "So, as soon as the Gun Club decision became known, representatives of Texas and Florida began pouring into Baltimore. From that day on, President Barbicane and other influential members of the Club were besieged day and night by formidable claims."[44] Victory for Florida leads to extensive economic development, transforming the near-wilderness of a fishing village called Tampa Town—population of 796 in 1870—into an industrial boomtown of 150,000, as seen in figure 3.1.[45] In *From the Earth to the Moon*, Verne developed, and transferred to his readers, much more than the technical requirements for spaceflight; he also outlined the economic, social, and organizational preconditions that he thought would be needed to make a voyage to the Moon a feasible proposition.

Although *From the Earth to the Moon* focuses predominantly on a single spaceflight, Verne also expresses, through Ardan, some of the sentiments that would later become part of the core philosophy of space expansionism. Although Barbicane is concerned principally with the practical challenges of the voyage at hand, Ardan ventures bolder statements about the future of space travel: "In spite of the opinions of certain narrow-minded people, who would shut up the human race upon this globe, as within some magic circle which it must never outstep, we shall one day travel to the moon, the planets, and the stars, with the same facility, rapidity, and certainty as we now make the voyage from Liverpool to New York!"[46]

TAMPA TOWN PREVIOUS TO THE UNDERTAKING.

TAMPA TOWN AFTER THE UNDERTAKING.

3.1 The economic development of Florida in *From the Earth to the Moon. Source:* Jules Verne, *From the Earth to the Moon Direct in Ninety-Seven Hours and Twenty-Minutes: And a Trip Round It*, ed. Louis Mercier and Eleanor King (New York, 1874), 66, 82.

Indeed, Ardan's self-proclaimed objective for the voyage is nothing less than lunar colonization, taking with him provisions for permanent habitation—including, of course, such vital necessities as cuttings from Médoc and Côte d'Or grapevines.[47] He requests annual follow-up resupply missions and foresees a future in which "later on trains of projectiles will be established between the earth and the moon!" This would allow for the company "to colonize the lunar regions, to cultivate them, people them, to take them all the wonders of art, science and industry!"[48]

At the conclusion of the story, with the triumphant return to Earth of the lunar travelers, Verne alludes to an even yet more adventurous spacefaring future for humanity to come: "Thus some time after the return of the travellers, the public received with marked favour the advertisement of a Joint-Stock Company (Limited), with a capital of a hundred million dollars, divided into a hundred thousand shares of a thousand dollars each, under the name of the National Company of Interstellar Communication."[49] At the conclusion of this carefully researched depiction of spaceflight, Verne describes the initiation of a for-profit company dedicated to interstellar travel, which receives funding at half of the relative level of the Apollo program—$100,000,000 in 1870, when *Around the Moon* was published, would be $371 billion dollars in 2024 GDP ratio equivalent terms.[50] With Verne and *From the Earth to the Moon*, the early seeds of space expansionism planted in earlier centuries had now matured into arguably the first flowering of popular space expansionist thought. It should therefore perhaps be unsurprising that this text would become the inspiration and stimulus of so many early spaceflight pioneers.

The depth and breadth with which Verne treated the technical, organizational, and social requirements of spaceflight was unparalleled. As a result, his novel had a profound impact. Verne described, for the first time, an almost technically plausible launch system and spacecraft design, with specific orbital calculations, underpinned by believable social and economic motivations, and laid it all out in a bestseller that was ultimately accompanied with compelling illustrations (see figure 3.2). After Verne, the transfer and theoretical development of spaceflight concepts, in a multitude of science fiction and nonfiction treatises, would begin to advance and proliferate markedly.[51] Whereas at the start of the century, a well-known man of letters like George Tucker felt the need to publish his speculative spaceflight fiction under a pseudonym, by 1900, even the US government's leading astronomer, Simon Newcomb, felt at liberty to publish openly a science fiction story dealing with the concept of the first spaceflight—which he thought, realistically and conservatively enough, would be a simple suborbital flight above the atmosphere rather than a grand voyage to the Moon.[52] By the end of the nineteenth century, a vast array of spaceflight concepts—from space stations to solar-electric-propulsion-propelled spacecraft to asteroid-mining colonies—had been suggested, developed, refined, and transmitted through works of spaceflight fiction and nonfiction treatises. Verne's *From the Earth to the Moon* had been foundational in both transforming the public's appetite for these types of stories and increasing their general believability.

The most consequential impact of Verne's work, however, was arguably not on subsequent spaceflight fiction but rather the direct inspiration that

3.2 Jules Verne's launch system and spacecraft for lunar orbit and return. *Source:* Jules Verne, *From the Earth to the Moon Direct in Ninety-Seven Hours and Twenty-Minutes: And a Trip Round It*, ed. Louis Mercier and Eleanor King (New York, 1874), 42, 122.

he provided to the three main pioneers of liquid fuel rocketry: Konstantin Tsiolkovsky in Russia, Robert Goddard in America, and Hermann Oberth in Austria-Hungary (today Romania). Tsiolkovsky explicitly attributed his early interest and passion for space travel to Verne: "I think the first seeds were sown by the imaginative tales of Jules Verne, which stimulated my mind. I was assailed by a sense of longing, and this set me to thinking in a specific way."[53] Hermann Oberth was engrossed by Verne's novel at the age of eleven, rereading the book five or six times, ultimately knowing it by heart, and crediting it with his lifetime devotion to making spaceflight a reality.[54] In an oral history interview in 1987, Oberth noted that his mother had copies of the German translations of *From the Earth to the Moon*, which she gave to him for Christmas in 1905 and which caused to him start thinking about spaceflight and, in particular, that "perhaps it would really be possible."[55] Although Goddard's writings suggest that H.

G. Wells's *The War of the Worlds* and its unauthorized American sequel, *Edison's Conquest of Mars*, were his most definitive inspirations, he also read Verne at a young age and was inspired by it. He even worked on his own unpublished version of *From the Earth to the Moon*, in which he corrected a number of Verne's technical errors and transferred the space travelers from a projectile fired from a cannon to a multistage rocket.[56] Verne's spaceflight story would continue to have an inspirational impact on the development of spaceflight over half a century after its initial publication, with Valentin Glushko, later the chief designer of Soviet rocket engines, writing: "In the spring of 1921 [at the age of thirteen], I read *From the Gun to the Moon* and then *Around the Moon*. These works of Jules Verne shocked me. While reading them, my breath was taken away, my heart was pounding, I was in a frenzy and was happy. It became clear that I must devote my entire life to the implementation of these wonderful flights."[57] *From the Earth to the Moon* thus not only transferred technical knowledge to the pioneers of spaceflight; it also fired their imaginations, fueled their passions, and inspired them to devote massive personal effort to pursuing in reality the dream that Verne had described in fiction. If the rocketeers of the early to mid twentieth century were the proverbial mothers and fathers of spaceflight, and Oberth, Tsiolkovsky, and Goddard the grandfathers, then Jules Verne is rightly the great-grandfather of spaceflight.

Verne continued to weave astronomical and space themes into his work throughout the rest of his long writing career. In *Aventures de Trois Russes et de Trois Anglais dans l'Afrique Australe* (*The Adventures of Three Russians and Three Englishmen in South Africa*) in 1872, he wrote about English and Russian astronomers working together to establish a meridian in South Africa; in *Le Pays de Fourrures* (*The Fur Country*), published the following year in 1873, he wrote about an astronomer joining a Hudson Bay Company expedition in Northern Canada in order to observe an eclipse of the Sun; and in *Hector Servadac*, from 1877, he imagined a group of European adventurers being swept up by a close-passing comet and proceeding to have an adventure traveling through the solar system before finally escaping the comet during its next flyby of Earth and traveling back to their home world in a balloon. *From the Earth to the Moon* would also continue to circulate in popular culture for decades to come, with the work first being produced for the stage in France as a four-act, twenty-three-scene play entitled *Voyage dans la lune* in 1875, and then, incredibly enough, as an opera in New York, in 1877, where it played at the Booth Theatre.[58] Georges Méliès's groundbreaking silent short film *Voyage dans la lune* in 1902 also drew direct inspiration from Verne, and, as the first motion picture about spaceflight, it inaugurated spaceflight's introduction into arguably its next most influential narrative medium.

Verne was also associated with another narrative that incorporated spaceflight and space expansionist ideas at their most extensive and expansive in a work that, in some respects, is almost as remarkable as *From the Earth to the Moon*: the short story "In the Year 2889." In 1885, Verne received a letter from the famous American newspaper publisher and exploration enthusiast James Gordon Bennet Jr., who asked Verne to write an imaginative story about America in the coming centuries. Bennett's primary notoriety in the history of exploration is that he was the sponsor of Henry Morgan Stanley's African expedition to find David Livingstone, which Bennett correctly assessed would provide exciting stories that would help sell more newspapers. He was also responsible for sponsoring the more tragic outcome of the USS *Jeannette*'s attempt to reach the North Pole. Regardless, the story that emerged in response to Bennet's request to Verne is unique and fascinating for several reasons, not least of which was that it was the only Verne story to have been published first in English—in 1889, in the American magazine *Forum* rather than Bennett's *New York Herald*. The extent to which Verne was directly involved in writing the story is debated, with the general view being that it was predominantly written by his son Michel. However, it was nonetheless published under the name of Jules Verne, when he was still an active writer, and it seems likely that the father and son duo at least collaborated on the outline of the narrative.

The story of "In the Year 2889" describes in broad brushstrokes a far-future civilization full of marvels: massive cities of tens of millions of people, temperature-controlled skyscrapers, new means of generating and storing vast amounts of energy called *accumulators*, and "aerial locomotion crossing the sky in all directions!"[59] The main character, such as he is, is Mr. Fritz Napoleon Smith, who purchases the old *Earth Chronicle* newspaper and then invents *telephonic journalism*, a system in which "instead of being printed, the Earth Chronicle is every morning spoken to subscribers, who, in interesting conversations with reporters, statesmen, and scientists, learn the news of the day."[60] This system has made Smith the wealthiest and most influential man in America, as evidenced by "the plenipotentiaries of all nations and our own ministers themselves crowding about his door, entreating his counsels begging for his approbation, imploring the aid of his all-powerful organ."[61] The story lightly covers a day in Smith's life and specifically covers spaceflight matters from the perspective of the "astronomical department" of the *Earth Chronicle*:

"We have phototelegrams from Mercury, Venus, and Mars."
"Are those from Mars of any interest?"
"Yes, indeed. There is a revolution in the Central Empire."
"And what of Jupiter?" asked Mr. Smith.
"Nothing as yet. We cannot quite understand their signals. Perhaps ours do not reach them."[62]

Here Verne—whether Jules, Michel, or some combination of both—imagines a future in which there are regular communications with a Martian empire and there are attempts to communicate with an inhabited Jupiter. Interestingly, the Moon is presented as being uninhabited and, curiously enough, unvisited—with the far side of the Moon treated as *terra incognita*. The work also holds one final prescient astronomical prediction: that the orbital elements of a newly discovered planet beyond Neptune, named *Olympus* in the story, had been worked out to a mean distance of over eleven billion miles and an orbit of over 1,311 years—remarkably not that far off from one of the largest trans-Neptunian dwarf planets that would ultimately be discovered in the early 2000s, Eris, which has an average distance to Earth of over six billion miles and an orbit of over 559 years.[63] This prescience is particularly noteworthy given that the discovery of the first trans-Neptunian object—the dwarf planet Pluto—was still some forty years away. Verne's imagination therefore clearly expanded far beyond the Moon and presaged not only our contemporary desire for interplanetary communications but our popular appetite for ongoing scientific discoveries about the scale of our solar system as well.

Verne's continued late-life enthusiasm for space narratives did not stop there. Verne only wrote one preface in his life for another writer's work, and it was for Irish writer Robert Cromie's story about a voyage to Mars, *A Plunge into Space*. Although *A Plunge into Space* had little of Verne's attention to technical detail, it was nonetheless popular, and it stayed in print from its first edition in 1890 until 1910. Verne wrote the preface to the second edition in 1891, which he addressed "To My English Readers." Although he noted he would "have preferred more details, more facts and figures in connection with the stupendous phenomena," he commended the book and its "weird and wild" voyage in "the Steel Globe."[64] The *Steel Globe* is the spacecraft in Cromie's novel, which works by way of an antigravity shield similar to Riddell's and which is built somewhere in the wilderness in Alaska by an English adventurer, Alexander MacGregor, and a Cambridge-educated engineer, Henry Barnett. Added to their crew—and it is arguably the first purposefully multidisciplinary spacecraft crew in fiction—are Sir George Sterling, "a financier in search of a new speculation"; Mr. Walter Durand, "a literary man in search of a new plot"; Mr. Victor Graves, "an artist in search of a new subject"; Mr. Charles Blake, "a politician in search of a post"; and Mr. Frederick Gordon, "a special correspondent in search of copy."[65] In a theme that would come to be more and more common, and which first appeared in Jules Verne's *Around the Moon*, as the crew returns from their successful mission to Mars, Sir George Sterling, the financier, plans to found a new space company: "I shall call it 'The Inter-Planetary Communication Company, Limited.' As to capital, I think a hundred millions would do to start with."[66] This amount is, of course, exactly the same as proposed by Verne's protagonists for their National Company of Interstellar Communication upon their lunar return. Already in the nineteenth century, then, the idea of very well-capitalized, privately financed space companies was being articulated as part of different spaceflight narratives—and already imaginary space company names were starting to revolve around a decidedly limited set of words and concepts.

Verne's last space-related work, *La Chasse au Météore* (*The Chase of the Golden Meteor*) was published posthumously in 1908 by his son Michel, who made significant additions and edits to his father's unfinished work. Nonetheless, it reveals a fascinating final insight into Verne's far-seeing astronomical imagination. The book is a low-key drama of competition between astronomers in Virginia, who claim credit for discovering a new asteroid that turns out to be made of pure gold, and thereby worth a fortune, and which threatens to change the geopolitical balance of power and wealth in the world when it crashes into a newly independent Greenland. Impressively, even here Verne precedes and predicts the future, with the newspaper printing breathless and credulous calculations of the value of this asteroid, estimating it at "not far short of six billions of dollars."[67] Calculating the current equivalent of this amount in 2024 GDP ratio terms, based on 1908 when it was published, this amount would be equivalent to $5.7 trillion dollars, giving Verne yet another science fiction first: the first to predict the journalistic enthusiasm for stories about "trillion dollar asteroids." He even predicted the entrepreneurial opportunity that some would see in this—in this case, in the form of a young, wealthy Parisian heir and technically inventive dilettante, the delightfully named Zephyrin Xirdal, who invents a radioactive ray emitter to change the orbit of the asteroid so that it falls to Earth, by which means he intends to claim ownership of all of its metal. Verne further imagines the popular concern that such a large amount of asteroidal metal might reduce global ore prices and fortunes, noting: "In general, the rich were in consternation, thinking of the probable depreciation of their fortunes. On the contrary, the poor were delighted by the prospect, fallacious as it might prove, of their having a share of the cake."[68] Ultimately, the competing claims of ownership—by the astronomers on account of their discovery, by Zephyrin Xirdal on account of his changing the asteroid's orbit, and by the Greenland State on account of the asteroid landing on its territory—are highly contentious, and so, in one of Verne's classically internationalist and high-minded flourishes, the protagonists decide to let the meteorite slip into the ocean, with the two competing astronomers choosing instead to save each other from falling through the ice during a recovery attempt, and becoming fast friends as a result. Although not a spaceflight narrative per se, as nobody in it ever leaves Earth, *The Chase of the Golden Meteor* nonetheless demonstrates the remarkable breadth of Verne's ability to combine astronomical and technical knowledge and to foresee the rather predictable motivations of humanity as it proceeds with its incremental engagement with the cosmos.

Throughout his life, Verne planted seeds for thinking about the practical considerations of spaceflight, the scientific value of astronomy, the potential for romantic adventures traveling on comets, interplanetary communications, and even the politics and economics of asteroid resources. As far-seeing as his many works proved, Verne's *From the Earth to the Moon* was unsurpassed in its influence on the future development of spaceflight. No other single work covered in this book would inspire more writers of spaceflight narratives or more spaceflight engineers. To this day, his ability to seemingly foresee that the first human voyage to the Moon would involve three people launching from Florida on a mission that would take around four days is remarkable. If there is one work that might be considered the platonic ideal of a spaceflight narrative, it is *From the Earth to the Moon*—a story that was written about a voyage into space that would occur in reality over a hundred years after the story was written, in part because the story itself encouraged and educated people over multiple generations to make it happen.

If *From the Earth to the Moon* is a central document in the historical development of the spaceflight narrative, then what can we say about its role in the evolution of space expansionism? It is sometimes claimed that much of the content in *From the Earth to the Moon* was meant to be satirical, and that we should be correspondingly cautious in interpreting the statements in the story to be representative of Verne's own thinking. This is, of course, always true when it comes to works of fiction, and the only spaceflight-related statements of Verne's known thus far are from his fictional works. As a popular writer, Verne also had a noted flair for the humorous, and many of the scenes and statements in the book were clearly written with Verne's tongue firmly in his cheek. At the same time, it should be said that Verne was not a famous satirist; rather, he was a popular adventure writer with a keen eye for the types of fantastical stories and narrative conceits that would both appeal to the public and have a basis in science and realism. Perhaps then it is more accurate to say that, though not necessarily a committed space expansionist himself, Verne—more than any other writer before him—was able to understand the popular appeal of space expansionist thought and rhetoric, and was able to imagine the potentially significant impact that its public articulation might have on motivating action and investment in the technologies and systems for spaceflight. It may be a step too far to classify Verne as a space expansionist, but that does not lessen his foundational impact on the development of space expansionism; even Karl Marx, after all, would famously declare that he wasn't a Marxist.[69]

Regardless, it seems that Verne's protagonist, Michel Ardan, probably should be classified as a space expansionist. Whereas Impey Barbicane initially seems satisfied with the glory and adventure of traveling to the

Moon, Ardan expresses an enthusiastic desire to thoroughly colonize the Moon and to transplant there the fullness of humanity: its arts, its sciences, its industry, its grapevines. Verne himself may or may not have fully subscribed to space expansionism as a worldview, but it seems clear that Michel Ardan most definitely would have. Verne portrayed the journey to the Moon as an archetypal nineteenth-century adventure of exploration and potential colonization, and so it is appropriate that he also presented arguably the first complete archetype of an enthusiastic space expansionist. In addition to Verne's remarkable ability to foresee many of the technical elements of a voyage to the Moon, he also seems to have acutely foreseen the particular type of character that would become so well-known to the global space communities of the twentieth and twenty-first centuries: the character of the space expansionist.

Verne and *From the Earth to the Moon* may be the epitome of the power of the spaceflight narrative, but there are still a great many ideas and concepts for spaceflight that Verne didn't cover. In particular, though Verne described how *spaceflight* might be achieved, he did not describe how humanity might realize long-term *space habitation*. For that narrative, we need to cross back over the Atlantic, to return to the shores of America—and specifically to the shores of New England—and dive into the story of how an American pastor, gunrunner, and social organizer wrote the first detailed description of how a group of humans would build a new home for themselves in orbit around Earth and begin to live "off world" for the first time.

4

A WORLD ON A BRICK SPACE STATION

Among the kind references to "The Brick Moon" which I have received from sympathetic friends, I now recall with the greatest pleasure one sent me by Mr. Asaph Hall, the distinguished astronomer of the National Observatory. In sending me the ephemeris of the two moons of Mars, which he revealed to this world of ours, he wrote, "The smaller of these moons is the veritable Brick Moon." That, in the moment of triumph for the greatest astronomical discovery of a generation, Dr. Hall should have time or thought to give to my little parable—this was praise indeed.

—Edward Everett Hale, preface to *The Brick Moon and Other Stories*, 1899[1]

By the late 1860s, not only had spaceflight narratives articulated a pseudo-plausible technical vision for human spaceflight around the Moon—but they had also articulated some of the technical, economic, and cultural dimensions of what today we would call *space expansionism* as an emergent worldview. In this regard, Edward Everett Hale stands as a giant—the first seer of the space station and one of the preeminent progenitors of American space expansionist thought. Understanding the story of Edward Everett Hale, and how he conceived of the opportunities and risks inherent in the expansion of humanity into the cosmos, can tell us much about the human processes that serve as the engines and motivations of spaceflight in the first place.

Edward Everett Hale was not a regular writer of science fiction or extravagant stories, as were Jules Verne and Edgar Allan Poe. Rather, he was a figure with a career and contribution to American public life far in excess of his literary contributions to spaceflight and space expansionist designs. In fact, he was one of the most influential American figures of the nineteenth century. Hale was an ardent promoter of "free state" emigration to the nation's western frontier, and a major fundraiser and organizer for the cause—in particular, as the Kansas Debate was heating up. He was the author of more than 150 books and articles, and he was a prolific magazine publisher and entrepreneur. Perhaps most interestingly, he was also a truly inspirational religious organizer and orator, in addition to being the father of a philanthropic social movement that at one time counted its branch organizations in the hundreds. For all these contributions to American public life, he was ultimately awarded high honors from the federal government in general and the US Congress in particular; he ended his life as chaplain of the US Senate, and he served as a prominent advocate for US participation in the Congress of Nations at The Hague. He was, indeed, that truly rare spirit—a man seemingly respected in equal measure by presidents and by the poor of his hometown of Boston.[2]

His most famous work, "The Man Without a Country"—published anonymously in 1863 under the name of his alter ego, Frederic Ingham—served as a major catalyst for the Union cause leading into the American Civil War, with its tale of an exiled traitor turned emphatic US patriot being remembered, even in 1880, as having "rivaled the 'Moon Hoax'" in its effect on the public.[3] In 1869 and 1870, the same pseudonymous Frederic Ingham would publish a serial narrative in the pages of *The Atlantic Monthly* with a similarly deft use of literary realism to tell the story of the first artificial satellite: "The Brick Moon." The story of the man-made moon and its thirty-seven inhabitants was well known and appreciated in its day, with it having achieved a sufficiently iconic status that when Asaph Hall discovered the two moons of Mars in 1877 at the US Naval Observatory in Washington, DC, he wrote to Hale and declared that the smaller of the two was "the veritable brick moon."[4]

Although its priority as the first work to describe an orbiting space station is well established, the content of "The Brick Moon" remains less well-known than it should be. The story was in fact the first American narrative to present spaceflight as a large-scale collaborative project requiring public engagement and fundraising. It also firmly established the relationship between spaceflight projects and utopian ideals, a relationship that would come to form the core of American space expansionism, which is all the more meaningful in light of Hale's almost forgotten role as an advocate for and active director of settlement and colonization efforts on the American frontier.

Although known principally for his literary and moral character, Hale's early life—growing up as part of one of the leading intellectual families of New England—was one with ample exposure to science and boyhood engineering.[5] His father—Nathan Hale Sr., an accomplished and experimental Boston printer—was an early advocate for railroads in America. Hale Sr. passed on his passion for the new technology to his son, born in 1822, who remembered watching the Granite Railway Company of Massachusetts, the first chartered American railway, as it carried granite to build the large civic monument to the Battle of Bunker Hill.[6] The elder Hale even set up a model railroad in the family parlor to show visitors just what was meant by the novel notion of a car running on rails.[7] As he matured, Edward and his older brother, Nathan Hale Jr., would build their own toy locomotives and engage in other activities that appealed to their interests in natural philosophy and the mechanical—experiments with chemistry sets, Leyden jars, telegraphic equipment, and climbing to the roof of their house in order to translate the semaphore messages from the lookout on Boston's Central Wharf.[8] The height of boyhood excitement came in 1833, when Hale was eleven years old and the Providence and Worcester Railroad reached Newton, seven miles from Boston. The Hale family made regular excursions to observe the operations, and, starting with the trial run of *The Meteor*, the first locomotive engine in New England, the eager and charming young Hale boys worked their way in with the engineers to participate in the early experimental journeys, in which they were "inducted into the mysteries of engine-driving."[9] With childhood experiences like these, the potential of technology and human industry must have seemed nearly limitless to the young Hale brothers. As such, it seems almost natural—although nonetheless incredible and wonderful—that the two brothers would, through their boyhood conversations, generate the first recorded concept of a space station.

As the first age of rail gathered steam in Boston, the young, imaginative boys of the Hale household entered Harvard College. There, amid the columns and courtyards, they began to think of other technical achievements that humankind might yet produce. Hale entered Harvard at the young age of thirteen, where he was fortunate enough to be taught mathematics by Benjamin Pierce—one of the most famous of early American mathematicians. At Harvard, he would organize his friends into what he called the *Octagon Club* to pursue a mutual enthusiasm for astronomy and to assist with meteor-counting research. Nonetheless, it was with his brother that he would continue to have his most thorough technical conversations.[10] The two Hale brothers discussed a wide range of youthful aspirations and schemes, one of which was evidently the idea of an artificial Earth satellite for aiding in the determination of longitude at sea, which was to be made of brick in order to survive the rapid, fiery ascent through the atmosphere.

In his preface to *The Brick Moon and Other Stories*, Hale revealed the origin of the idea: "The account of the first plan of the moon is a sketch, as accurate as was needed, of the old chat and dreams, plans and jokes, of our college days, before he left Cambridge in 1838"—with "he" here referring to his brother Nathan.[11] In a later letter to his younger brother, Charles, Hale also referred to the project jocularly as "our great enterprise."[12] Hale's story of "The Brick Moon" thus seems to have been much more than a flight of fancy. It seems to have been a dramatized account and fictional realization of an idea that he and his brothers, and possibly others, had in fact seriously discussed. This fact alone deserves to be more widely appreciated and celebrated, and it is arguably a space story worthy of its own dramatization in film: the true story of the young college students who first imagined a global positioning satellite system for the world in the 1830s.

That a project for an artificial satellite was being discussed with some seriousness in the dining halls and dorm rooms of 1830s Harvard indicates how increasingly close the region beyond Earth's atmosphere was beginning to seem to some nineteenth-century Americans. It was not, however, until Hale had mulled the idea over for some further thirty years, and had considered the sociological implications of living on an artificial satellite of Earth, that he put the idea to paper.

"The Brick Moon" begins with Hale's alter ego, Frederic Ingham, recounting the old college conversation that ultimately leads to the construction of the brick moon:

We came across this business of the longitude, and, as we talked, in the gloom and glamour of the old South Middle dinning-hall, we had going the usual number of student's stories about rewards offered by the Board of Longitude for discoveries in that matter. . . . Like all boys, we had tried our hands at perpetual motion. For me, I was sure I could square the circle, if they would give me chalk enough. But as to this business of the longitude, it was reserved for Q, to make the happy hit and to explain it to the rest of us.[13]

Q—a reference to the middle name of Hale's beloved older brother—then employs the concept of a peashooter to explain the idea of a launch into space to his younger brother and his friends:

"If from the surface of the earth, by a gigantic pea-shooter, you could shoot a pea upward from Greenwich, aimed northward as well as upward; if you drove it so fast and far that when its power of ascent was exhausted, and it began to fall, it should clear the earth, and pass outside the North Pole; if you had given it sufficient power to get it half around the earth without touching, that pea would clear the earth forever. If only we could see that pea as it revolved in that convenient orbit [i.e., a polar-orbiting satellite on a specific meridian], then we could measure the longitude from that as soon as we knew how high the orbit was, as well as if it were the ring of Saturn."

"But a pea is so small!"

"Yes," said Q., "but we must make a large pea."[14]

The young students then set to work to design a satellite that would be very large and very light—with their eventual solution being a two-hundred-foot hollow brick sphere supported internally by seven smaller hollow brick spheres. A constellation of two or four satellites is even considered to allow for easier observation, but it is agreed that they could begin with one. As for the launch itself: "It was not to be by any of your sudden explosions. It was to be done as all great things are done, by the gradual and silent accumulation of power."[15] A giant flywheel battery—built of oak and pine and hooped with iron—would store up the energy generated from a waterwheel until the energy was sufficient to spring the flywheel and send the brick moon into orbit. As they sketched out their plan, they realized that the twelve million bricks required for the moon would alone cost $60,000 ($19,300,000 in 2024 PWC ratio equivalent terms)—an unimaginable sum for young men with college bills still unpaid. With this realization, the idea is reluctantly shelved. It is impossible to know how closely Hale's description of this fictional Harvard College discussion mirrored the actual Harvard College conversation he experienced with his brother. However, given Hale's lifelong habit of using experiences and characters from his own life to give his stories verisimilitude, and the fact that the simple plan was fully within the conceptual reach of bright, young American minds in the 1830s, it seems probable that this record of the discussion includes at least as much fact as fiction. The first recorded concept of utilizing an orbiting artificial satellite to aid in global position determination can therefore most likely be attributed to a couple of precocious American teenagers.

Unlike Hale and his college friends, however, the fictional Ingham and his fellow dreamers reunite after seventeen years to put their plan into action. One of the friends, George Orcutt, had become a civil engineer and eventually grown rich and powerful as a railway baron.[16] Another, Ben Brannan, had no wealth but had become a firebrand orator of the first rate and was renowned as a trustworthy man of impeccable honesty and sagacity. Hale's brother, Q, is only peripherally involved with the story from this point on; his inclusion in the tale serving mainly to credit his brother with the initial idea. The other three, however, had never let their youthful idea stray far from their minds:

> And now, George was ready to consecrate a round hundred thousand to the building of the Moon; and Brannan was ready, in the thousand ways in which wise men move the people to and fro, to persuade them to give to us a hundred thousand more; and George had come to ask me if I were not ready to undertake with them the final great effort, of which our old calculations were the embryo. For this I was now to contribute the mathematical certainty and the lore borrowed from naval science, which should blossom and bear fruit when the Brick Moon was snapped like a cherry from the ways on which it was built, was launched into the air by power gathered from a thousand freshets, and, poised at last in its own pre-calculated region of the ether, should begin its course of eternal blessings in one unchanging meridian![17]

Thus appears the first reference to a satellite being "launched" in the English language.

Hale's description of the fundraising campaign for the brick moon shows a remarkable attention to the details and requirements of large, public subscription campaigns. Throughout, the massive project—initially estimated at $200,000 ($732,000,000 in 2024 GDP ratio equivalent terms; $64,500,000 in 2024 PWC ratio equivalent terms)—is portrayed as a civic endeavor voluntarily undertaken by a self-selected group of citizens without government involvement. The trio of leaders appeal to philanthropy and the practical need for a better method of determining longitude—noting that "the horrible loss of the *Royal Martyr*, whose longitude was three degrees awry, startled the world, and gave us a point to start from."[18] Hale suffuses the story with the type of tactical details that he would have been familiar with as an experienced fundraiser himself. First Ingham explains to the wealthy Orcutt that "he must not subscribe the one hundred thousand dollars in a moment. It must come in bits, when 'the cause' needed a stimulus, or the public needed encouragement."[19] They then "caught neophyte editors, and explained to them enough to make them think the Moon was wellnigh their own invention and their own thunder."[20] Next they sent invitations for a "private meeting" at Orcutt's parlors to three thousand men of science, philanthropy, and commerce, an invitation that attracted only a smattering of press, a friend of Brannan's, and three men, "all adventurers whose projects had failed."[21]

The prewarmed press, however, nonetheless gave "first-rate notices" of the gathering, and the trio organized for their first public meeting a few days later. There, now with a larger crowd and in a room with the walls covered "with intelligible and unintelligible diagrams," the three gentlemen made their pitch—with Ingham as an old Navy captain first explaining the need for a better method of determining longitude, with Orcutt the civil engineer and man of industry vouching for the project's feasibility, and with Brannan the evangelist left to rouse the crowd to enthusiasm until "each man felt as if he were himself the inventor, who had bridged difficulty after difficulty; as if, indeed, the whole were too simple to be called difficult or complicated."[22] The presentation is rounded out with Ingham underlining the philanthropic, civic nature of the project: "We could not go forward without the general assistance of the community. It was not an enterprise which the government could be asked to favor. It was not an enterprise which would yield one penny of profit to any human being. We had therefore, purely on the ground of its benefit to mankind, brought it before an assembly of Boston men and women."[23] Contemporary space advocates and space entrepreneurs may find themselves delighted, and perhaps a little embarrassed, by the many similarities between Hale's description and the techniques used to promote nascent space projects today.

With the campaign initiated, the supporters begin to come forward in the idiosyncratic manner of actual public subscription projects. After the presentation, Thomas Coram, "prince of merchants," who had lost a ship to a bad chronometer, stands and asks for the estimated cost. Ingham replies that $214,729 would be required for the first moon, with subsequent moons costing $159,732. Coram then suggests that $250,000 will ultimately be required for the first moon and offers that he would be "one of twenty men to furnish the money," with the funds to be returned to the subscribers if the moon is not finished within twenty years.[24] Orcutt then indicates, as planned, that he would be another of the twenty, before a third man, visiting from England and moved by the spirit of the moment, announces that he would be a third. Some of the ladies in attendance were moved to organize a Great Longitude Fair to raise support, while the core trio hit the streets with their subscription book to raise funds, leveraging a wide array of motivations: those who subscribe because they were certain the amount would never be raised and wanted "to get the credit of liberality for nothing"; those who did not care about longitude, or particularly believe in the project itself, but personally believed in Ingham, Orcutt, or Brannan; those who subscribed because "it was the thing to do, and they must not be out of fashion"; and those who subscribed because they remembered the day they heard that a ship with a friend or a loved one had been lost.[25] Even with all this, however, they only counted $162,000 in subscriptions and could not use a cent of it, except Orcutt's, until the whole amount had been raised. Feeling that "the orange was squeezed dry" and that no further public subscriptions could be expected, they all returned—for the moment—to their daily lives.

This detailed and engaging account of a public subscription campaign resembles closely the fundraising histories of early astronomical observatories—most notably, the American civic observatories in Cincinnati, Albany, and Detroit, including the initially failed campaigns that sought to establish the Harvard Observatory.[26] It is perhaps more likely, however, that Hale drew his observations and inspiration from a source closer to home—one of deep relevance to his role as one of the forefathers of American space expansionism: his personal experience as director and president of the New England Emigration Aid Company.

The New England Emigration Aid Company was a pioneering organization in its effective promotion and assistance of western settlement efforts. When the delicate political balance of the Missouri Compromise was threatened by the imminent passage of the Kansas-Nebraska Bill, Eli Thayer, a Worcester schoolmaster and member of the Massachusetts Legislature, became determined to ensure that the new territories would become free states by organizing concerted Northern emigration to Kansas and Nebraska. Although granted a charter for the company in 1845, he found few allies until he spoke publicly at Worcester, where Hale—serving there as Unitarian minister of his first parish—became, as Thayer remembered it, "the very first man to express confidence in [the Company's] success, and his own readiness to work for it with all his might."[27] Hale immediately sent Thayer a copy of his own tract, *How to Conquer Texas Before Texas Conquers Us*, which similarly advocated for the emigration of nonslaveholding citizens to the newly added state of Texas.

Hale soon began passionately advocating in public speeches and in the press for the cause of the new company, which would arrange low-cost transportation for companies of settlers and provide financing to establish critical infrastructure like sawmills, gristmills, and printers in each new town settlement. Hale became one of the company's principal directors, and later its president. He also compiled a 256-page handbook for emigrants, and even assisted with gunrunning efforts for the Northern colonists when it seemed that the settlers might have to defend their rights with arms after five thousand nonresidents from Missouri rode over the border into Kansas in order to swing territorial elections in favor of a governor that was more sympathetic to the South. Hale's principal role, however, was as a fundraiser. The original 1845 charter had grandly declared it would raise a fund of up to $5,000,000—equivalent to an astonishing $77 billion in 2024 GDP ratio equivalent terms, and an almost equally incredible $3.2 billion in 2024 PWC ratio equivalent terms. Although the company never reached this massive sum, in eight years of active operation its expenditures totaled an impressive $136,000 (approximately $1.6 billion in 2024 GDP ratio equivalent terms, $78.6 million in 2024 PWC ratio equivalent terms), with Hale claiming that this allowed as many as five thousand settlers to move west with the company's assistance.[28] The contribution of the company was even greater than the number of settlers assisted, however, creating as it did a business model for assisting western emigration that was copied extensively and that provided "the backing of social and monetary prestige" for "the unknown pioneers at the front."[29]

Hale would remain intellectually and emotionally committed to the idea of western settlement throughout his years. In 1891, he would preach a sermon on "the colonization of the desert" after traveling by rail across the country to California, and he would declare, near the end of his life: "I am more proud of my part in the settlement of Kansas . . . than I am of any public service I have ever rendered."[30] That the first American description of the financing and development of an artificial satellite came from one of the most effective fundraisers and supporters of western frontier settlement underlines the fact that genuine conceptual links were starting to be forged in nineteenth-century America among spaceflight, finance, and the sociopolitical opportunity of frontier settlements. American western expansion and the origins of American space expansionism are therefore, at least through one prominent individual, genuinely entwined.

Before Hale can engage in his exploration of the sociological aspects of life on the space frontier in "The Brick Moon," however, his alter ego and his companions must finish their fundraising and build the moon, and in this too, Hale draws on his personal experience with fundraising—this time, related to his role as a promoter of government war bonds in the Civil War. After no further subscriptions seem forthcoming and the project is put on hold, Ingham decides to set out on a new path: He founds a new parish church in rural district number 9 of the Third Range of Maine. There, much to his delight, he finds nearby an unclaimed plot with a ready supply of yellow clay for brickmaking next to a river running down the hills and surrounded by ample timber for the proposed dam and flywheels. With new low-cost sources for their inputs, Ingham contacts Orcutt and the two refigure their plan, making new "estimates for feeding Irishmen, estimates for hay for mules . . . estimates for cement, estimates for the preliminary saw-mills, estimates for rail for the little brickroad, for wheels, for spikes, and for cutting ties," resulting in a new total cost estimate of $197,327.[31] Although they are still short $40,000, Orcutt secures the land, and he uses the funds to proceed with building the dam and flywheel to build investor confidence. As the dam and flywheel system begins to accumulate energy, the attack on Fort Sumter begins, an event which Ingham and Orcutt initially believe will doom their project but which in fact provides the final funds.

During the Civil War, Hale had thrown himself into support of the Union effort, promoting recruitment, drilling the Salignac Rifle Corps, serving as a civilian aid to General Benjamin Franklin Butler, and tirelessly contributing to the Sanitary Commission, army hospitals, the care of prisoners and refugees, and the education of freedmen.[32] He had also advocated for citizens to provide loans to the Northern state governments. The fictional Ingham followed the same path, and it was through this financial instrument that the final funds of the project are raised—with the $59,000 in hand for the project being invested in these loans, which fortuitously provide a return of "more than 50%" by the end of the war.[33] The subscribers are then told that the total funds had been raised and that the construction of the brick moon can begin in earnest.

The construction of Earth's first artificial satellite in rural Maine is described as a large project, with over a hundred laborers and requiring a project-management structure with a main foreman and separate divisions of labor for accounting, the brick kilns, the carpenters, and the commissariat. The work proceeds faster than planned—as "the men had no chances at liquor or dissipation" and the new brick machines "made brick inconceivably fast"—but the work nonetheless continues into the winter. With the change of the seasons, Orcutt, Brannan, the heads of the divisions, and their families all move into the seven hollow spheres that existed within the structure of the brick moon to shield themselves from the winter weather. While Ingham is away in Boston trying to raise a further $25,000 in order to paint the moon with a feldspathic granite dust that would fuse into a white enamel during the fiery ascent—which would increase the brick moon's visibility—the raging winter river cuts away at the river banks on which the flywheel and brick moon are positioned. This triggers a premature launch while the winter inhabitants of the brick moon are asleep inside. Although the brick moon was not designed to be a first home for humans in space, it thus accidently becomes one.

Ingham initially assumes that tragedy has struck and that his friends and their families have burned to death in the ascent, or suffocated once they went beyond the atmosphere. After almost a year passes, however, he notices in the *Astronomische Nachrichten* that a new asteroid has been discovered with an enormous movement in declination, so much so that Ingham thinks it might be the brick moon. He then proceeds to buy the largest Alvan Clark telescope he can afford to search for it, and he incorporates the report of a second newly discovered asteroid with unusual movement as part of a scientific search for the brick moon that would anticipate the calculations and observations of satellite watchers in the century to come:

> We calculated in the next week sixty-seven orbits on the supposition of so many different distances from our surface. I laid out on a paper, which we stuck up on the wall opposite, the formula, and then one woman and one man attacked each set of elements, each having the Logarithmic Tables, and so in a week's working-time, the sixty-seven orbits were completed. Sixty-seven possible places for [the brick moon] to be in on the forthcoming Friday evening. Of these sixty-seven, forty-one were observable above our horizon that night.
>
> She was not in one of the forty-one, nor near it.[34]

Although it takes a significant number of observations, the brick moon is eventually found in orbit 5,000 miles from the surface, and Ingham goes in search of a large telescope to allow him to observe it more clearly.

Hale makes a striking comment on nineteenth-century American observatories when he describes Ingham's search turning up an unused observatory with a large refractor. He begins by noting that the fictional Tamworth Observatory in the story had been unutilized for ten or so years, and provides, as explanation, a brief history of the observatory that is a pastiche of some of the American observatories of the day: "The trustees had quarreled with the director, or the funds had given out, or the director had been shot at the head of his division,—one of those accidents had happened which will happen even in observatories which have fifteen-inch equatorials; and so the equatorial here had been left as useless as a cannon whose metal had been strained or its reputation stained in an experiment. The observatory at Tamworth, dedicated with such enthusiasm,—'another lighthouse in the skies,'—had been, so long as I have said, worthless to the world."[35] Hale was thus acutely aware of the popular enthusiasm for astronomy in the nineteenth century, and he was evidently a close follower of astronomical developments: referencing the leading international astronomical journal of the day, name-dropping the most famous American telescope maker, quoting fellow Bostonian John Quincy Adams's description of astronomical observatories as lighthouses in the skies, and being intimately familiar with the human foibles that often caused the utility of nineteenth-century astronomical observatories to be short-lived. "The Brick Moon" is thus very much a spaceflight narrative steeped in the astronomy-based space exploration culture of its day.

The Tamworth Observatory fulfills its role for Ingham, allowing him to see the brick moon with enough detail to make out figures on its surface and to interpret the Morse code messages they are trying to send by making, as a group, spectacular, low-gravity-enabled jumps—a high jump for a dash, a low jump for a dot. The message received, Ingham coordinates with one of the main supporters of the project to return the message with large black letters made of cambric on the sawmill flat near the launch site, where the supporter has built his own large reflecting telescope to aid in the search. With communications established between the ground and the thirty-seven inhabitants of the artificial moon, Hale can begin to explore the question that had motivated him to write the story in the first place: the question of social development in extreme isolation from the Earth.

The three serialized stories of the "The Brick Moon" were written principally to establish the background for the fourth, the separately titled "Life in the Brick Moon," in which Hale links, for the first time, the idea of utopia with orbital space habitation.[36] At the end of the first three stories, the principal mission of the brick moon project has failed, as the accidental launch did not insert the satellite along a specific meridian and the resultant drift in its orbit negates the project's use as a determinant of longitude. With the addition of life and human society to the object, however, the brick moon acquires a new purpose as an extreme environment for the development of a community. Although the notion that the two-hundred-meter sphere has enough mass to maintain an atmosphere—and that the weather-worn empty brick spheres accidentally held enough plant matter, animal life, soil, and ice to allow for agriculture and livestock husbandry—is decidedly implausible, the narrative has at this point moved beyond its emphasis on verisimilitude in order to focus on more philosophical matters. Indeed, the earlier efforts to establish a plausible story of fundraising and construction were affected principally to induce the nineteenth-century reader to genuinely consider the idea of a community in orbit around the Earth. As Hale's son would note in his biography of his father, "The Brick Moon" was certainly "one of the most outrageous" of his father's stories, but it was nonetheless "based on a real sociological conviction."[37]

What, then, was the sociological conviction that required Hale to develop a community on an artificial moon in space in order for it to be expressed? His conviction was of the possibility of an attainable utopia—what Hale referred to as "the New Civilization."[38] Although Hale would develop his concept of utopia elsewhere—such as *Sybaris and Other Homes*, and in his emphatic lobbying for the idea of a world court that, he believed, could one day settle national disputes and bring peace to the world—the idea of an isolated life in Earth's orbit allowed him to highlight some of the most salient aspects of his utopian vision. The brick moon had on its surface three acres of land for exercise, with the main habitation area underground—ample enough room for thirty-seven people, especially compared with the cramped quarters of many people in nineteenth-century Boston.[39] With almost perpetual sunlight at five thousand miles above Earth, Hale describes the residents of the brick moon as being able to achieve six, eight, or ten harvests in a year, as well as enjoying an almost Elysian climate. As Orcutt messages to Earth: "I can see your storms. We have none. When we want to change climate we can walk in less than a minute from midsummer to the depth of winter. But in the inside we have eleven different temperatures, which do not change."[40] With a near-perpetual harvest and de facto climate control on the brick moon, Hale's description bears a striking resemblance to the language and vision used over a century later by Gerard O'Neill and the space habitation movement of the 1970s and 1980s. Thus, even in the mid-nineteenth century, space expansionist thinking was becoming developed enough to imagine and articulate the idea of manufacturing of an artificial world that could be superior to the Earth.

The core of Hale's vision for the brick moon, however, was the superior form of society that would be enabled—in fact, almost forced—by the conditions of life on a space station. For one, they have less responsibilities and "this decrease of responsibility must make them as light-hearted as the loss of gravitation makes them light-bodied."[41] He even alludes to a positively Edenic existence in their freedom from the constraints of fashion, and perhaps clothing entirely, with Ingham commenting, "I did not say that I believed they wore fig-leaves in B.M., but that was my private impression."[42] The population of the brick moon also conducts the kinds of social experiments, such communal living, that many proponents of space colonies in the 1970s would be similarly eager for: "We tried hotel life, and we like it, for there has never been the first quarrel here. You can't quarrel here, where you are never sick, never tired, and need not be ever hungry. But we were satisfied that it was nicer for the children, and for all round, to live separately, and come together at parties, to church, at signal time, and so on."[43] There is an almost complete freedom from labor, with "three hours head-work" being the maximum allowed. Most of the day is taken up with meals, prayers, conversation, and evening activities such as singing, lectures, theater, and dancing, until curfew sounds at ten and everyone is in bed by midnight.[44] So effortless and naturally pleasant is life in orbit portrayed to be that "the only law on the statue-books is that every one shall sleep nine hours out of every twenty-four."[45]

In addition to the utopian elements of life in space, the question that most interests Hale about the brick moon is the question of personal and societal independence. Hale describes how the extreme isolation and low population size of life in Earth's orbit would lead to an independence of thought and spirit. The subtitle of the last episode in the brick moon saga is "Independence," and it is the idea of separation from the world—and the potential benefits that would result—that dominates the conclusion of the story. The members of the brick moon, along with Ingham, are all Sandemanian—a Christian religious minority that originated in Scotland and was the target of a good deal of ill will from Unitarians and others at the time of Hale's writing. As described by Hale, the independence of life in Earth's orbit allows for this religious minority to happily establish their own Sandemanian church and continue to conduct their lives. Those still on Earth try to interest the inhabitants of the brick moon in terrestrial affairs, and in a schism within the Sandemanian church in particular. The orbital inhabitants give no response to this, however, and take negligible interest in the affairs of the world, responding with disdain when it is suggested to them that they should establish a civil government: "Go to thunder with your old law books. We have not had a primary meeting nor a justice court since we have been here, and, D.V. [*deo volente*, or god willing], we never will have."[46] Hale uses orbital spaceflight as a tool for exploring what constitutes "the good life"—and whether life, removed from Earth and centered in space, might, at least hypothetically, be an improvement. In so doing, Hale introduces the question of social values to orbit—and how social values in space might end up different than social values on Earth.

Here then is the idea of the new start, the break with the world and with history that would come to animate American space expansionism—the idea of a small, isolated space community that has achieved full productivity and happiness with no further need of Earth and which is then free to develop independently under its own rules. Indeed, as Hale describes it, it is the very smallness and isolation of the brick moon that has allowed it to achieve utopia. With a small population, the tight-knit community is a natural development, and it greatly eases the difficulty of "instruction and civilizing and of evangelizing in general."[47] With isolation from the world, the complexity of life is reduced and thus happiness more easily attained without the broader political concerns of the terrestrial world. A new type of society, more agreeable, at least to Hale, thus becomes possible through the confined, isolated nature of life in space. In the final passage of "The Brick Moon," Hale eloquently steers his readers to consider whether life might be better under these conditions: "Can it be possible that all the human sympathies can thrive, and all human power be exercised, and all human joys increase, if we live with all our might with the thirty or forty people next to us, telegraphing kindly to all other people, to be sure? Can it be possible that our passion for large cities, and large parties, and large theaters, and large churches, develops no faith nor hope nor love which would not find aliment and exercise in a little 'world of our own'?"[48] Increase the number of people by a factor of twenty or thirty and this description comes close to one that might be expressed by a hypothetical O'Neillian colonist on Island One in the classic twentieth-century space expansionist text *The High Frontier: Human Colonies in Space*.[49]

With his realistic depiction of the organizational and engineering aspects of spaceflight, and his genuine hopes for an attainable utopia in orbit, Edward Everett Hale's spaceflight fiction reads like an American combination of Jules Verne and H. G. Wells's later efforts. Although it would not be as influential as *From the Earth to the Moon,* "The Brick Moon" is unequaled in the range of its contributions to the early American conceptual development of spaceflight and American space expansionist thought. It was spurred by a real conversation at Harvard in the 1830s concerning the implementation of a global positioning satellite constellation. It described a small group of willful and capable men returning in adult life to fulfill their boyhood dreams of space. It provided a realistic description of a public financing campaign for a satellite that could easily be mistaken for that of a modern space start-up, and it presented the first hopeful thematic combination of spaceflight and orbital utopia, a narrative combination that would come to dominate popular space expansionist visions in the century to come. This combination, from the pen of an avid proponent and organizer of frontier American settlement, was truly unprecedented.

Where then do "The Brick Moon" and "Life in the Brick Moon" stand within the intellectual history of space expansionism? Unlike *From the Earth to Moon,* there is little evidence to suggest that Hale's narrative significantly inspired the early pioneers of spaceflight. Like Verne, it would be too far to say that Hale himself should be considered a space expansionist given that his expression of space expansionist ideas and sentiments occurred entirely in fiction and did not seem to form a significant part of his personal philosophy. However, whereas *From the Earth to the Moon* is primarily a scientific adventure story, the philosophical consideration of the possibilities and constraints of life in orbit is the central purpose of Hale's spaceflight narrative. Although others had written utopian narratives of life on the Moon, prior to Hale's work there had been no writing at all about life on an artificial, human-made world in orbit around the Earth. Hale described the motivation for constructing that first space station in utilitarian terms—the creation a global positing satellite system to solve the longitude problem—but the ultimate use of the station was for experiments in utopian living. "The Brick Moon" therefore may have been less influential than *From the Earth to the Moon,* but it was also significantly more philosophical. Even if we cannot say that Hale was a space expansionist, the depth of his consideration of both spaceflight and practical frontier settlement suggests that he should probably at least be classified as genuinely space expansion curious, and there is a case to be made that

Hale should be considered the first American philosopher of space expansionism and the progenitor of its core interests in orbital utopias.

Although Hale had only tentatively extended an imagined expansion of life into space some five thousand miles from the surface of the Earth, over the next three decades others would extend it throughout the entire solar system. Edward Everett Hale embodied the connections that were then forming between the exploration of the heavens, utopian idealism, and the expansion of the nation westward. He would hardly be the last to do so, for what followed Hale would be a veritable explosion of American space expansion narratives.

5

AMERICAN CONQUEST OF THE SOLAR SYSTEM

Monarch of Fate is man above all destiny. Man yet shall chain the stars; shall drive the harnessed worlds.

—Florence Carpenter Dieudonné, *Rondah; or, Thirty-Three Years in a Star*, 1887[1]

In the last quarter of the nineteenth century, a wild narrative literary force emerged from the United States that laid out a bold, exciting, and at times disturbing space future: a space future of space colonies, space mines, middle-class lunar vacations, billionaire missions to the Jovian moons, and conflict and conquest across the solar system. Five books demonstrate this development particularly well: *The Princess of the Moon* by Cora Semmes Ives; *Bellona's Husband* by William Roe; the anonymous *Man Abroad*; Florence Carpenter Dieudonné's *Rondah*; and John Jacob Astor's *Journey into Space*. In reading their narratives and their imagined space futures, we can understand better the origins of our own present, and perhaps potential future, space realities. In these stories, the roots of many of our contemporary space visions can be found. By the end of the nineteenth century, through these narratives, the grandest of cosmic fantasies had been articulated, including at scales of economic and social interplanetary organization that far exceed anything we are even close to today.

As these new American concepts for spaceflight developed, and as the nation's knowledge of engineering and industry grew to preeminence in the world, following on from Edward Everett Hale's vision, spaceflight began to be conceived of as the result of large-scale engineering projects—in one case, solar system–wide engineering projects—with America presented as the wellspring of these efforts. As an increasing number of Americans began to consider spaceflight a natural result of advancing technological progress, by the end of the century the solar system was being presented as an opportunity for continued economic expansion, adventure, and social experimentation in the tradition of the American frontier. By the end of the century, the American space expansionist narrative had even enthralled its first American billionaire.

Although the focus here is on the development of American space narratives in the late nineteenth century, there was one truly remarkable and revolutionary Scottish-Canadian space narrative from the mid-nineteenth century that was written before any of the other narratives in this chapter. William Leitch was born in 1814 on the Isle of Bute, near Glasgow in Scotland. He would become a minister, astronomy popularizer, and wandering lecturer—including lecturing publicly in favor of the plurality of worlds hypothesis—before emigrating to the United Province of Canada in 1860 to become the principal of Queen's College (known today as Queen's University, in Kingston, Ontario).[2] While serving as the principal at Queen's, he published an essay entitled "A Journey Through Space" that articulated, with remarkably forthright simplicity, for the first time in a nonfiction context, the idea that rockets could serve to power the exploration of the solar system: "The only machine, independent of the atmosphere, we can conceive of, would be one on the principle of the rocket. The rocket rises in the air, not from the resistance offered by the atmosphere to its fiery stream, but from the internal reaction. The velocity would, indeed, be greater in a vacuum than in the atmosphere, and could we dispense with the comfort of breathing air, we might, with such a machine, transcend the boundaries of our globe, and visit other orbs."[3] This remarkable statement appeared first in the Scottish magazine *Good Words* in September 1861, and it would be reprinted in Leitch's only significant book, *God's Glory in the Heavens*, which presented a tour of the solar system on a comet, but with the potential for rocketry to allow humans to voyage to other planets identified clearly, if only briefly, right at the start.

Leitch's spaceflight narrative, like many that came before it, is one that primarily identifies astronomy and space exploration as methods to enable a fuller appreciation of the handiwork of God. Today it is primarily his pioneering nonfiction reference to spaceflight rocketry that is of greatest interest, but his own focus was on drawing attention to the worlds of the cosmos as a pathway to engage with the divine:

> In the survey we mean to take of the heavens as illustrative of God's glory, we shall first direct our attention to the moon, our nearest neighbour. The moon will form the first step in the ladder by which we shall attempt to scale those heights from which we may command the widest range of the marvelous works of the Almighty. Although we cannot by searching find out God, although we are baffled in our attempts to comprehend the Absolute, still there are stepping stones across the abyss of space, which enable us to enlarge our view, and to form a juster conception of the Infinite and the Eternal.[4]

Amid his religious paeans to the infinite, however, the book also includes a number of far-seeing predictions and insights. Leitch impressively anticipated that there would be rings around Neptune, that traveling at the speed of light would be needed to travel to the stars, and that asteroids could potentially be hollowed out and used by space colonists: "The very globe itself might be tunnelled and split up, so that contending parties might have little worlds of their own to live in. The imagination can thus easily revel in the wildest fancies, if we exchange the normal conditions of life for extreme physical suppositions."[5] Despite his prescient insights, Leitch's clear and prescient prediction of the ability of rockets to enable space travel did not seem to have much, if any, impact on his contemporaries, and his contribution was essentially either unknown or forgotten in space-related literature until its rediscovery in the 2010s. Nonetheless, he foresaw the ability of rockets to enable cosmic journeys more clearly than anybody else would for many decades. With this preeminent Canadian spaceflight narrative accounted for, we now can return to the more complex, if somewhat less technically accurate, American spaceflight narratives of the nineteenth century.

As might be expected when examining books from the last half of the nineteenth century in America, we unfortunately begin with a work that makes for uncomfortable reading: *The Princess of the Moon*, subtitled *A*

Confederate Fairy Story. This slim publication from 1869 does not feature any new technical concepts for spaceflight; nonetheless, it is the earliest example of a spaceflight narrative by an American woman yet found, and it remains a striking example of both the enduring projection of utopian yearnings on the Moon and a reminder that one person's utopia can sometimes be another's dystopia. Published in Warrenton, Virginia, it states on its front cover that it was "Written by a Lady of Warrenton, Va." The "Lady of Warrenton" in question was Cora Matilda Ives, née Semmes. She was married to Colonel Joseph Christmas Ives, who in 1857 and 1858 commanded an expedition to explore the Colorado River by sternwheel steamboat, later served as an engineer and architect for the Washington National Monument, and joined the Confederate Army in 1861. Despite being born in New York City, Ives rose through the ranks to become aide-de-camp to Confederate President Jefferson Davis from 1863 to 1865. The Iveses lived in Richmond, Virginia, where they socialized with senior Confederate leaders and where Cora would stage elaborate theatrical performances to entertain the foreign dignitaries and journalists that Davis would regularly bring by their home. After the war, Cora moved to New York with her husband, but she clearly yearned for the "good old days" of the antebellum South, as *The Princess of the Moon* makes abundantly clear.

The opening line gives one a sense of the content and tone of much of the book: "It was after the dreadful struggle between North and South that a poor Confederate soldier, Randolph, wandered in a wood near the ruins of his once splendid home."[6] Randolph, after mournfully gazing on his ruined home, looks up at the Moon and sighs, "Ah! Had I wings, how gladly would I seek shelter on your distant, peaceful shores, sweet moon."[7] At this, the Fairy of the Moon instantly appears in a long veil of silver gauze and a blue velvet robe embroidered with stars and expresses her desire to help the soldier, who promptly asks her to "transport him to her dominions, as life in a conquered land had become to him a burden." The Fairy of the Moon gives Randolph a magical winged horse named Hope, who can transport him to the Moon. Before doing so, however, he surveys from the air some twenty thousand people who had been driven from Atlanta, and another twenty thousand dwelling in the ruins of Columbia, South Carolina. He then surveys "the homes of the conquerors, abounding in plenty, and decked in the spoils they had so cruelly acquired."[8] In disgust at what he sees, Randolph commands Hope to "bear him at once to his destination, and in a short time he arrived safely in the moon."[9]

This initial method of lunar transportation in *The Princess of the Moon* was equivalent to the lunar spirit journey of George Fowler's similarly named *Randalthus* of 1813. The Moon itself is described in the manner of a utopian fantasy. Ives's moon is ruled by a king "both wise and good" and is populated by people that have never even heard of war. The king lives in a magnificent palace, which is home to a beautiful princess who is moved by Randolph's plight and whose courtship is the focus of most of the book's content. Randolph makes numerous journeys on Hope back to Earth to visit with the fairy, as well as to the very gate of Heaven, before ultimately being granted permission by the king and the fairy to marry the princess, which he does while dressed in "a magnificent Confederate uniform."[10]

It is during its dramatic wedding scene that the book becomes interesting from the perspective of its spaceflight narrative. The wedding of Randolph and the unnamed princess is interrupted by members of "a meddling, peddling race, called 'Yankees'" who arrive on the Moon via several balloons, which begins both the most interesting and disturbing section for the modern reader. The Yankees shortly begin congratulating themselves on "discovering a new country," which they believed would perhaps surpass "the best government the world ever saw." They set at once to begin a survey of the Moon, with the Yankee balloon-borne lunar explorers almost ignoring the existing lunar inhabitants.[11] The Yankee explorer-invaders then proceed into the palace, where they are understandably shocked to see a Confederate soldier on the lunar throne. Ives then introduces a racist trope—that of the faithful slave—into this lunar fantasy, depicting an unnamed man who calls Randolph "Massa" and who had "heard that an expedition was forming to survey the moon, and having dreamed that he should find me [Randolph] here, he was induced to join the party in hopes of realizing his wishes."[12] The Yankees also turn out to be "the very persons" who burnt down his home, at the discovery of which they try to escape, but the fairy turns their balloons into dragons. They are only saved by Randolph's clemency as he muses that "life would become insupportable if fellow creatures do not assist and encourage each other."[13] Randolph then asks the Yankees to repent their ways as he allows them to return by balloon to Earth, asking them to "unshackle the race of heroes they have enslaved," referring to the South without a hint of irony.[14] The story ends with Ives suggesting a future Yankee invasion of the Moon to come, noting that after returning to Central Park in New York, the lunar balloonists "have never again undertaken to invade, but there is no knowing what may yet take place, as they are a very indefatigable people."[15]

The Princess of the Moon, despite—or rather through—its politics and ideology, is a relevant example of the ways in which spaceflight and space expansionist ideas were beginning to seep into popular culture. While the Confederate Randolph travels to the Moon by the romantic grace of a fairy and a white-winged horse, the Yankees travel by balloon and are focused on surveying the land for future development, and possible future invasion. The idea that human invention could enable lunar access and territorial acquisition was incorporated into a story that was otherwise a literal fairy tale. This is also arguably the earliest spaceflight narrative to depict a human expedition to the Moon as a negative development, in that the Yankee balloon-borne lunar explorers are depicted as predominantly avaricious and domineering in their activities and intent. *The Princess of the Moon* also seems to be the first spaceflight narrative that features two different human expeditions to the Moon, each with their own conflicting priorities, as well as a projection of terrestrial human conflict on to the lunar surface for the first time. This distinctly American lunar fairy tale is thus emblematic of its time—and a timeless reminder of the subjectivity of "utopian" projections on the Moon, or anywhere else in the cosmos for that matter.

In December 1880, a more uplifting spirit, one of promise and hope in American industry, returned to the American spaceflight narrative with Washington Gladden, the father of the Social Gospel movement and a charter member of the American Economic Association. Gladden penned a short juvenile about a future visit to the Moon on a passenger liner—a 150-foot-long, electric-powered "great aërial car" called the *Meteor*, the supposed result of a new invention by Thomas Edison, who only the year before had introduced his light bulb to the public. Gladden's work includes a number of noteworthy elements, including a striking drawing of the massive great aerial car (see figure 5.1); the need for humans to wear oxygen masks on the Moon due to its weak atmosphere; near-mute lunarians due to the same; the notion that the new "aerial line" led to lunarian emigrants

5.1 An early spaceflight fiction lunar liner, *The Meteor*. *Source:* Washington Gladden, "A Christmas Dinner with the Man in the Moon," in *Santa Claus on a Lark and Other Christmas Stories* (New York, 1890), 27–49.

arriving in London; the first use of the term *moonquake*; and an entreaty to the reader at the end of the story that he or she "must remember that in these days of Edison and Jules Verne, nothing is incredible."[16] Although the story of the *Meteor* marks Thomas Edison's first appearance as a brilliant inventor in a spaceflight fiction narrative, he would reappear in the American spaceflight imaginary some nineteen years later, even more powerfully, in *Edison's Conquest of Mars*, a work that would serve as one of Robert Goddard's core inspirations. Gladden's "A Christmas Dinner with the Man in the Moon" is also notable as one of the first stories to describe voyages to the Moon as a regular, commercial occurrence and as an experience that was accessible to middle-class passengers, much in the manner of train journeys, which had become dramatically more accessible by the end of the nineteenth century. It is perhaps fitting that the idea of a trip to the Moon that might be affordable for ticket-buying members of the public would be introduced in the American spaceflight narrative by one of the leading forerunners of the progressive movement, who had argued throughout his career for greater labor rights, expanded social welfare, and desegregation.

In 1884, a second American spaceflight narrative by a female author was published, Polly Cabell's "A Trip to the Moon."[17] The story has been attributed to Mary Hopkins "Polly" Cabell (1769–1858), who was born into a distinguished Virginia family and married John Breckinridge of Kentucky, a senator and attorney general in Thomas Jefferson's cabinet. The tale attributed to her was published in an obscure Louisville, Kentucky, magazine for young readers called *Electra: A Belles Lettres Monthly for Young People*. However, this would have been some thirty years after her death, which raises the possibility that the story may have been wrongly attributed to her and was perhaps instead authored by one of her children or grandchildren (one of her grandchildren was also named Mary, Mary Cabell Breckinridge, though there is no evidence that she used the name Polly.) Whatever the correct attribution, the story recounts how a Dutchman, while preparing a giant cask of beer for a forthcoming drinking festival, is hurled into space when the cask explodes. He reaches the Moon, landing safely in a snowdrift on the slope of an extinct volcano. He finds a barren, uninhabited Moon—appropriately so, in accord with the scientific understanding of the time—and he bemoans his fate as he faces the prospect of death by starvation and regrets ever having encountered the fateful beer cask. But then he awakens to find himself safely back in his bed on Earth and realizes that the explosion had simply knocked him out and sent him into a spaceflight dream. Ms. Cabell ends the story with a pun, saying that, fortunately for him, "his keg of beer was not his bier after all." The temperance movement was on the rise at the time of the story's publication, spearheaded by the foundation of the Women's Christian Temperance Union in 1873, and with this closing line it seems likely that the motivation for the story was in service at least partly of this movement, particularly given that the Breckinridge-Cabell family were strong Presbyterians. For those interested in spaceflight thought, however, it is Ms. Cabell's opening observation that is the most interesting and prescient: "We are not moon-stricken, but moon-strikers, or endeavoring to be such. But how can we strike the moon except in imagination? Well, then, let us deal with imaginary things as if they were realities, for imaginations often become realities."[18] This prescient insight into the potentially self-fulfilling nature of imaginative spaceflight narratives, and the expression of Americans as a people eager to strike out for the Moon, makes Cabell's short story, as whimsical as it is, without a doubt the most interesting story of a lunar voyage to come out of the American temperance movement.

As the 1880s progressed, the publication rate of American spaceflight fiction increased, the utopian strains of the American spaceflight narrative proliferated, and a number of far-thinking writers articulated an ambition to become not only Moon strikers, but Mars and Venus strikers as well.[19] As astronomical observations made it increasingly clear that the Moon did not have a significant atmosphere, and thereby had limited hope for life, the projection of utopian alien civilizations moved further afield to the planets. The Moon is very large and visible in the night sky, to the point where its different features—most notably its mare—are visible to the naked eye. The planets, however, are so far away that they are often difficult to easily distinguish from stars unless an individual has a trained eye or recourse to knowing the positions of the planets through astronomical tables or other means. They are therefore significantly more remote, both in reality and in familiarity, to the imagination than the Moon. Interestingly, this seems to result in imagined alien societies on the planets being frequently depicted in the latter half of the nineteenth century as being more advanced than those on Earth, whereas for the Moon in earlier eras the assumption was just as often the reverse. One example of this is the self-published work "The Key of Industrial Co-operative Government," written in 1886 by Henry Francis Allen, an idealistic Swedenborgian sign painter in St. Louis. In a somewhat awkward combination of doggerel poetry and social analysis, Allen expressed his desire to be a Venus striker along with the hope that the "machinery and intelligence of the age" might create a utopia on Earth equivalent to that imagined to exist on Venus, whose fictional civilization had already mastered faster-than-light travel.[20] Already then, by the 1880s, with hopes for faster-than-light travel, and regular, commercial passenger cars to the Moon, the aspirations of American spaceflight narratives were beginning to far outstrip those of earlier eras.

In 1887, under the name Hudor Genone, West Point graduate Lieutenant William James Roe published his tale of Mars striking: *Bellona's Husband,* the first known spaceflight narrative from an American business entrepreneur.[21] Roe, who had lost his family's moderate fortune through bad investments and had been president of the failed Hydrogen Company of New York, tells a tale seemingly at least somewhat inspired by his own experience. The story is of an unwitting New York investor who invests $5,000 ($10,800,000 in 2024 GDP ratio terms; $1,450,000 in 2024 PWC ratio terms) into a venture involving an "ethereal disc" made of solidified hydrogen and who becomes president of the company. Although the ethereal disc successfully transports its inventor and the investor to Mars, the proposed venture, which is the first suggestion of a point-to-point suborbital spaceflight transportation company, is never realized, and the investment is lost.[22] The opening to *Bellona's Husband* nonetheless includes some of the first explicit discussion of space-related business models and space investment in fiction.

Professor Garrett, the inventor of the ethereal disc in the book, declares: "A capitalist . . . is proverbially timid. Let me once persuade our moneyed men that my enterprise is feasible in a commercial way, and the alacrity with which they will take hold and push it would astonish you. It is a low view of human nature; but, after all, it is the pocket that must be appealed to."[23] Professor Garrett then proceeds to lay out the first suborbital point-to-point business model: "I construct a disc of suitable size, say at Calcutta, load it with rice, pepper, indigo, cashmere shawls, and so forth. We launch our craft, rise above the influence of gravity, adjust our armature, and in about eleven hours descend in the neighborhood of New York."[24] Roe, who evidently had significant personal insight into the psychology of wishful and ambitious investors, also provides the first description of the thought processes of a fair-weather space investor:

> I must say that the project appeared to me very alluring. It was so original and new, just one of those grand ideas that burst upon the world like a thunderbolt. Then, too, the profit! No doubt, after the matter had been demonstrated, the machines could be sold at an enormous profit. Besides, there was the glory. How proud I should be to be pointed out in the street as the far-seeing capitalist who had been the means of enabling the professor to bring his invention before the public! And how proud Amelia would be, and how it would delight the children

> to have their father spoken of as—in some sense—an illustrious man! In earlier life dreams of fame had been mine . . . The previous year my head had been examined, and the person who operated had given me a chart, wherein my most prominent characteristic had been set down as imagination.[25]

One would hardly be surprised to hear these sentiments expressed, if perhaps only privately, by today's space investors and venture capitalists.

This story by a failed technology entrepreneur, which later transitions from describing terrestrial spaceflight business models to a fantastical romance adventure among a reverse-aging population of English-speaking Martians, represents the saltation that occurred in the conceptual development of spaceflight in the last decades of the nineteenth century. In an era that introduced new wonders like electric lighting and liquefied oxygen to the world, even something as spectacular as spaceflight seemed like it might be only a modest capital investment away. The nation had entered a new era of futuristic thinking and techno-optimism, best represented by the 1887 publication of Edward Bellamy's epochal *Looking Backward: 2007–1887*, which, in its poetic flourishes of cosmic sentiment, announced: "The long and weary winter of the race is ended. Its summer has begun. Humanity has burst the chrysalis. The heavens are before it."[26] It was in this era of rising American optimism and utopianism that the first work emerged that can legitimately be said to describe—however crudely—the economic development of the solar system.

In 1887, a small book was published in New York that, for the first time, described in detail an expansive vision of a future in space that remains far beyond the achievements of modern space activities even in the twenty-first century. The anonymous author of *Man Abroad: A Yarn of Some Other Century* conceived of the solar system as nothing less than the battleground of a vast political economy, with independent human populations on Venus, Mars, Mercury, Jupiter, Saturn, the Moon, and the asteroids—their initial colonies having been seeded from the Earth—all being in competition with each other, and with the home world, for raw materials and political power.[27] Almost a century and a half before *The Expanse*, the baroque politics of *Man Abroad* could allow it to be classified as the first space opera, if its interests weren't so firmly focused on human activity in the aggregate—being more interested in presenting an analysis of the federal system and world governance, planetary colonization, and political experimentation on the asteroid frontier than on telling the stories of individual politicians, colonists, or frontiersmen. *Man Abroad* can also be read as a vision of the late nineteenth-century globalization process playing out on a grand scale—complete with interplanetary trade and labor mobility.[28] Of all the forgotten works of nineteenth-century American spaceflight fiction, *Man Abroad* stands out as the most readable and interesting to the modern reader—with its conceptual achievements ranging from the first suggestion of human terraforming of other worlds to the first discussion of willful space settlement for political and social experimentation. Throughout, *Man Abroad* presents a distinctly American perspective on the movement of humanity into space—with the United States of America encompassing the entirety of planet Earth at the beginning of the book and encompassing the entirety of the solar system by the end.

Man Abroad begins in the office of the president of the United States, the government of which now encompasses the entirety of Earth. The president attends to the regular business of state—new ambassadorial nominations, the concerns of congressional districts, and commercial treaties—except that here the ambassadors are the United States representatives to Venus, Jupiter, Saturn, and Uranus; the congressional districts include the Transvaal, Brazil, and Calcutta; and the commercial treaty is one that is being negotiated with the people of the Moon. A reporter, Webster Lockmore—one of the only named characters in the book, the principals generally being referred to as "The President of the United States," "The Finance Minister of Venus," "The Premier of Jupiter," and so on—is in the Oval Office to interview the president about his plans for civil service reform, which the president states is required on account of the fact that "our territory now embraces the whole world, and may in time include other worlds."[29] Indeed, it is this extension of American manifest destiny into the solar system—specifically, to the Moon—that drives the thin narrative of the book. The Parliament of Jupiter, concerned that American expansion to the Moon would endanger their own "similar purposes of gradual and indefinite extension," decides to preempt the issue and declares war on the United States with "a fleet of electric ethereal ships" carrying an army of one hundred thousand men to the Earth-Moon system.[30] This scene predates Garrett Serviss's *Edison's Conquest of Mars* by over a decade, which is more commonly identified as containing "the first-ever scenes of massed fleets of interplanetary spaceships."[31] Soon, as the scramble for resources and interplanetary power gains momentum, Venus and Mercury are drawn into the conflict, until, exhausted by war and threatened by America's (i.e., Earth's) overwhelming forces, all of the planets of the solar system apply to become members of the United States.

The thrust of *Man Abroad* is not the story itself but its description of politics and interplanetary affairs, and in this it excels, being as it is the first work to describe in detail that core tenet of space expansionism: the settlement of the solar system. Although the process is described through a curious conceit of a discussion between ghosts on an asteroid, and the technical ideas for spaceflight described in the novel are crude at best, the language and description bear a striking resemblance to that of twentieth-century space settlement promoters. The discussion begins with the observation that "hardly a day passes but an aerial and ethereal car lights on one of these asteroids and colonizes it with human beings . . . look how many they have settled and developed, Jupiter, Mars, Venus, Saturn and the Moon."[32] The spectral conversation acknowledges the skepticism that had accompanied the idea of space settlement but which had been overcome by humanity's simple animal tenacity: "When I was alive on Earth, they used to tell me that a man couldn't live in the Moon or on any other planet; but he is doing so, all the same. He accommodates himself to his environment, just as the oysters and monkeys did before him. He is fruitful, and multiplies, and he is gradually taking possession of the solar system."[33] Twenty-eight years after the first publication of *On the Origin of Species*, space settlement already had adopted a Darwinian tone.

With the idea established that the solar system is undergoing widespread economic development, the discussion turns to the origins of this process. As *Man Abroad*'s future history has it, the movement into space began "just after the Americans conquered the rest of the world that was worth conquering," with the affair being a uniquely American endeavor, as "nobody else would have thought of it."[34] Indeed, it was American industry and innovation that enabled this interplanetary migration, with an American inventor having "improved the ordinary electric air-car" to allow it to travel into space; thereafter, "in a few months the over-crowded plains of Earth were losing population in the thousands."[35] Although the planets were recognized to be inhospitable worlds, it was discovered that "the electricity" of the ether was so strong that it could be "brought into play upon the Moon's surface to make oxygen, nitrogen and carbonic acid gas, by developing, in proper proportions, the rapid cultivation of plants and herbs that give off these gases."[36] This scientifically informed description of the process of terraforming other planets is many decades ahead of its time, with the term *terraforming* itself not being coined until 1942 in Jack Williamson's novella "Collision Orbit," which he published under the name Will Stewart.[37] With the process of terraforming begun, the settlement of the solar system proceeds in much the same manner as space advocates would envision decades later: "The Moon was rapidly settled up, then Venus, then Mars, then Jupiter, then Saturn, and now the asteroids."[38] Reference is made to the settlement of Uranus and Neptune as well. Twenty-five years after the passing of the Homestead Act, the gradual and indefinite expansion of the American people had already been projected to one day encompass all known planets.

Man Abroad also introduces another core tenet of American space expansionism: the notion of the space frontier as a crucible for political and social experimentation. Although Hale had alluded to this theme, his space station inhabitants and their experiments were accidental ones, whereas the anonymous author of *Man Abroad* describes a group of political idealists moving to the asteroid belt in order to implement the ideas of the most famous American political economist of the time. The introduction of this asteroid utopia comes from the journalist Webster Lockmore, who, while traveling between the planets, runs into ether-car trouble and is required to stop at a repair station on the obscure asteroid world of Henrygeorgia: "A peculiar people inhabited this asteroid. There was but little communication between them and the United States, and the Jupiterns and the Venusians had even less to do with them, while the inhabitants of Mercury and Mars ignored them entirely. The prevailing notion was that the population of the asteroid consisted of Socialists."[39] As he awaits his repairs, Lockmore enters into conversation with an intelligent asteroid farmer of "Yankee ancestry," who emphatically denies the accusation of socialism and declares that he and his people "are Individualists of the most pronounced type."[40] The farmer then relays the history of his ancestors who had colonized the asteroid in order to implement the utopian vision of Henry George.

Henry George was one of the most prominent progressive American political philosophers of his age. His best-selling *Progress and Poverty: An Inquiry into the Cause of Industrial Depressions*, first published in 1879, had turned him into a household name and laid out his plan for a more egalitarian and productive socioeconomic system. Impressively, George had even articulated an early form of the "Spaceship Earth" concept that would become popular in the Space Age, referring to Earth as a ship "on which we sail through space": "It is a well-provisioned ship, this on which we sail through space. If the bread and beef above decks seem to grow scarce, we but open a hatch and there is a new supply, of which before we never dreamed. And very great command over the services of others comes to those who as the hatches are opened are permitted to say, 'This is mine!'"[41]

In *Progress and Poverty*, George argued that a principal cause of income inequality is that landowners and monopolists capture a large portion of the wealth created by technological improvement due to their ability to charge, without taxation, for access to scarce natural resources, while productive activity is taxed heavily. Based on this premise, he advocated that a single tax be levied on landowners, based on the unimproved value of the land: a simple policy that would, to his mind, lead to a more efficient and productive society. The first colonists of Henrygeorgia had moved to the asteroid to implement this economic system. Interestingly, though the anonymous author of *Man Abroad* is evidently sympathetic to the political philosophy of Henry George, the initial implementation of the system by the asteroid colonists is described as a failure. The result of the single tax was large surplus revenues for the government and, as a result, "Socialism came, and Individualism went."[42] Subsequent generations of colonists then set about to reform the system, setting the minimum taxable plot of land at ten acres and taxing only unimproved and poorly maintained land so as to encourage productive improvement and "the sense of independence stimulated by land holding," while at the same time limiting government revenues and discouraging landlordism, squatterism, and shantyism.[43] The result of these alterations to the Georgist platform are a veritable space utopia on the asteroid, where "sewer and water and gas connections are furnished at low rates—in some cities for nothing" and the "laws are so respected that they administer themselves."[44]

It is not the political economy of Henrygeorgia per se that is remarkable, but rather the simple fact that the solar system was now being depicted in the late 1880s as within the domain of consideration for the evolution of human politics and economics. There is even a cursory questioning of the applicability of terrestrial economic systems under the rather different circumstances in space, with Lockmore asking what point there would be in applying an antimonopoly principle to land where there was access to the almost limitless worlds of the solar system. The response is as much political and ideological as it is economic: "Certainly there is more land than can ever be monopolized, if Henrygeorgia's principles were applied everywhere. But they are not. A syndicate of your United States Senators can go and lock up a continent in Neptune any day. Every member of your Senate owns an asteroid now . . . they may yet lock up the whole Solar System against the poor man who emigrates from earth, and the squatter on a planet may be told to move on into the depths of space."[45] The solar system of *Man Abroad* may be governed by realpolitik and baser human nature, but there nonetheless remains a hope for a utopia among the planets based on the ideals of independence and hard work—ideals that were evidently as important to American space expansionist thinking in the 1880s as they would be in the 1980s.

Man Abroad had clearly articulated the idea that space habitation could purposefully enable the development of new socioeconomic paradigms. The asteroids represented a new frontier for political experiments, and, just as importantly, the encounter with the space frontier led to alterations in the initial sociopolitical vision, resulting in a new synthesis that ultimately does bring about a utopian world on an asteroid. Almost a full century before Dandrige Cole and Gerard O'Neill would look to the asteroids with similar ideas of social experimentation and economic development, the planetary bodies of the solar system were already beginning to be portrayed as soil in which to root the potential extension of America's continuous political experimentation into the heavens.

With *Man Abroad*'s precocious expression of American space expansionist ideals, it is worth noting that the social and political landscape of the solar system that it describes is one that is governed almost entirely by private action and motivations. The fierce individualism of the colonists of Henrygeorgia is a prime example of this elevation of the individual agent and the denigration of government action, with the intelligent farmer noting that "governmental powers should not be given the benefit of any doubt whatsoever; the presumption being that the Government had no right to do anything that could possibly be done without its aid."[46] In a similar vein, the plans for the American conquest of the Moon are the result of "a secret organization of private citizens" whose power and connections are extensive enough to influence the action of the government.[47] Although a few of the Venusian colonists had been exiled by American expansion, the majority had reached the planet through voluntary emigration, depicted as an independent and personal process that was core to movements across the solar system—perhaps the most intimate reflection of which is the description of a Jupitern prison guard pining for his "stock farm" in the asteroids. The anonymous author of *Man Abroad* even goes so far as to depict the intrinsic motivations of the wealthy as the rightful guide to America's expansion into the solar system, with the president selecting Much Tin of Peking—the vice president of the Earth-Mars Ether Fast Line—as the best man to handle the treaty with the Moon because he is "rich enough to be beyond temptation."[48] Humanity's migration out of the cradle of Earth is not portrayed as the result of government action, prestige, or profit, but rather as the result of individuals—from the American inventor of the ether-car to the independent Yankee colonists of Henrygeorgia—each governed by their personal motivations and a human desire for the independence of land ownership and the freedom to use that land to live a life of their own choosing. Although it may seem that it is only recently that space has become a domain for the projection of the ideals of rugged individualism and libertarianism, *Man Abroad* shows that the planets and the asteroids have held a narrative appeal for their projection for over a century prior to their rise to prominence within spaceflight communities in the late twentieth and twenty-first centuries.

The ideas of cosmic independence and individual freedom also feature prominently in *Rondah; or, Thirty-Three Years in a Star*, a groundbreaking work published in 1887 by one of the pioneering women of speculative fiction, Florence Carpenter Dieudonné.[49] Relatively little is known about Dieudonné, and what little is known comes predominantly from the short entry for her in *A Woman of the Century: Fourteen Hundred-Seventy Biographical Sketches* published in 1893.[50] Dieudonné was born in Stockbridge Falls,

New York, in 1850. Her family then moved to Wisconsin, where she finished her education. After marrying, she spent time in Minnesota, then traveled through Europe and ended up in Washington, DC, where she was president of the Parzelia Club, "a conversational and literary order," and a member of the Woman's National Press Association, before dying in San Francisco in 1927. She published several poems and works in her life, though *Rondah* remains her most well-known work, in large part due to its reputation as a proto–space opera and the truly unusual conception of an alien world that she presented.

Rondah has an eclectic structure, presented by multiple narrators, and an even more eclectic narrative. The method of travel into space is unusual as well, with the four main characters hitching a ride on "a circle of meteoric worlds and fragments" in an elongated orbit that once every thirty-three years intersects the orbit of earth, at which point the "small planets mingled with masses of chaos are brought almost to the tops of our mountains."[51] The meteoric worlds are reached during their closest approach to Earth by an elaborate spacecraft launching device developed by a hermit in the Adirondack Mountains: "a spiral car which, by means of subterranean cavities, walled with cement and filled with certain explosives of his own combination, should be shot into the coming meteor. All the machinery of the contrivance is hidden and connected with a secret clock. At the precise second when the worlds strike right the clock will explode the torpedoes and the car will be launched in space!"[52] The obsessive and inventive hermit who created the device, Gregg Dempster, had never used it, in part because he miscalculated the orbit of the meteor world by not accounting for how the planets would change its orbit. The main protagonist, Regan Farmington, had befriended Dempster and thereby learned of the device with dramatic results: Dempster "shared his great hope, which had become the nerve of his life, with another man and from the lessened tension died."[53]

The primary underlying theme of *Rondah* is of Christian triumphalism and salvation through cosmic encounters, in a manner reminiscent of *An Account of Count D'Artois and his Friend's Passage to the Moon*, though with a story that is altogether more dramatic, strange, and expansive. A wilderness preacher, Father Renaudin, accompanies Regan to the meteoric world and declares that "for man to migrate from star to star is an impossibility!" Regan replies with a far-reaching statement of human ambition in the cosmos, which seemingly even predicts the subjugation by man of the stars themselves: "Monarch of Fate is man, above all destiny. Man yet shall chain the stars; shall drive the harnessed worlds!"[54] This phrase embodies the message and philosophy of *Rondah*, with it serving as both the opening and closing sentences of the book. It also presages the heightened and grandiose nature of the American space expansionist ambitions that would follow in the twentieth and twenty-first centuries. For Dieudonné, this cosmically expansive ambition is entwined with her interpretation of the promise of Christianity. Shortly after Regan's declaration, a disembodied voice responds to his bold statement with a resounding "And I will give him the morning star!"—a phrase from Revelation 2:28 that serves as a dual reference to both the planet Venus and the promise of Christ. At the end of the book, the eponymous Rondah, Regan's love interest, saves the meteoric world—which she names Parzelia—through her own kind acts and sacrifices. Parzelia then enters the orbit of Jupiter as a permanent moon, where she and Regan can serve as eternal masters of their new world in perpetuity.

Rondah has been lauded within science fiction critique for being one of the first to defend "alien rights"; however, humans are presented throughout the narrative as the superior species, and the supposed sympathy for alien rights does not stop Regan from killing and beheading a number of the inhabitants of Parzelia.[55] Shortly upon landing on the world, Regan unfurls "a large silken banner wrought with a showy design in red and gold" and declares, "I claim this world, with all its contents, by right of discovery! It is my just possession! Its sovereign am I! Its seas are my ways, its lands my domains, its people my subjects!"[56] Dieudonné seems to show some ambiguity about this assertion, with Roy Lee, the first narrator of the book, objecting to Regan that "you will not found a monarchy where no such absurd abomination has ever existed!" and declaring that Regan is "a scoundrel, a fiendish monster! See how he toys with lives and souls."[57] Nonetheless, by the end of the book, Regan is literally ascendant and, together with Rondah, the ruler of Parzelia.

Rondah is nonetheless remarkable for the early examples it provides of genuinely alien creatures and a truly alien world. The first alien encountered is a monster, one as large as twenty elephants and with "a head like that of a hippopotamus on a huge trunk which looked like a tree bole, a great flattened body like a turtle's, a black shell which we had thought lava, long paws ending . . . in human-shaped hands."[58] After the heroes escape the clutches of the giant hippopotamus-turtle monster, the meteoric world goes through a twenty-year winter. As they explore the world, the heroes remark upon its unusual characteristics: "Here were trees with silver fringes, others with crimson leaves, and masses of shrubs whose leaves were the purple velvet of the pansy. The landscape became like a beautiful dream; its coloring was of a kind which paled the hues of Earth to insipidity."[59] The world is full of volcanoes, a green-hued atmosphere, and other novel planetary vistas that have more in common with the multicolored pulp fiction visions of the early twentieth century than those of the late nineteenth. Even if *Rondah* may seem backward-looking in its embrace of a sort of interplanetary divine right of kings, it is nonetheless deeply forward-looking in its conceptualization of a truly alien planet, complete with multidecade seasons dictated by its extended elliptical orbit and a thoroughly diverse array of alien biology.

The people of the meteoric world are also unique: both bird-like in being able to fly and also decidedly vegetable in origin, being born in pods. Upon encountering this new intelligent species, however, the response by the travelers from Earth focuses on how they can be used to advance their goals, rather than recognizing any intrinsic worth in their own unique alien lives. On encountering them, Roy Lee enthuses that they had discovered "thousands of subjects for Regan; thousands of artisans for me! Sailors for air ships, toilers at looms!"[60] They soon set about to "civilize" the vegetable-bird people, and they soon have "a great army of docile, winged servants" who seem to serve as a sort of idealized subordinate species: "There appeared to exist no envy, hate or anger among them. They were ambitious, but they were not treacherous. Their faith in us, in all we hoped and in all we did, was perfect."[61] The subtext of species dominance and submission becomes explicit when Roy Lee notes that the vegetable bird-people: "With all their brilliance and intellect, lacked something. They would never be a dominant race. They must have leaders, and I am compelled to acknowledge that in Regan they had found one."[62] Regan proceeds to cut off the wings of some of the vegetable-bird people, believing them to be "hideous deformities."[63] This kills them, prompting Roy Lee to condemn the act, noting that "calling men vegetables does not change the crime! . . . These creatures have souls! They are the humanity of the sphere!"[64] Despite the clearly expressed belief in the superiority of humans over these alien beings, here is the sympathy for alien rights with which *Rondah* has been credited. That sentiment, however, is undercut shortly after when Roy Lee also admits of Regan, "He was our law. Why not kill ten men? Now and then in some such disagreeable way he made us all realize his power."[65] In the end, although *Rondah* introduces the idea that extremely different types of alien intelligences could be considered to have equality under human value frameworks, when it comes to the governance of a new planet, it also ultimately sides with individual authority and power.

At the end of the book, with Regan and Rondah living on Parzelia, which is now in orbit around Jupiter, they discover an ancient "Pre-Adamite" library. There they find, in dialogue with an ancient spirit, the grand timescale from which humans had emerged, initially from interstellar origins:

> In time primeval, before the sun with its attendant worlds was spoken into being, where now we whirl, on this tiny orb, there rolled two stupendous, peopled worlds more vast than Jupiter. On these dwelt such as I, men whose lives were prolonged for ages. We reasoned for cycles. We built for great periods. We had years granted until we learned too much and grew weary of our knowledge. But woe and sin and hate, intensified in evil according to the size of the field and the capacity of the race, existed with us . . . I noted the signs of impending disaster. I even heard the creaking of the sinews of our old world, and saw the continents being deserted for life beyond; but I was so busy, I hastened to get to a few more nations, and before I was there, while I was saying, "Some other time I will lay down my life," the globes, like bubbles in air, burst into millions of fragments and my existence was indefinitely prolonged.[66]

This grand scale of human cosmic narrative—a story of a humanity that began on another world around another star, prior to a great sundering planetary cataclysm—is arguably *Rondah*'s most interesting contribution to the evolution of the spaceflight narrative. Although some of these elements are found in other, earlier stories, it is in *Rondah* that the prose begins to stretch out in time and scope to prefigure the stupendous narrative scale of Olaf Stapledon's *Last and First Men* and *Star Maker* some fifty years later. The pre-Adamite spirit that watches human history unfold—the rise and fall of both Atlantis and of ancient Egypt—relays to Regan a hope that humanity can yet build a new and better future in the stars: "As the star floats you shall teach it all things; make better laws than Earth dreamed of; create a paradise! Man should learn to lay each stone to stand for a thousand years. Your records should be on walls of amber, not in books of silver."[67] The American spaceflight narrative thus begins to encompass not just the solar system and its planets, but also the stars, and even time itself.

In a penultimate flourish, the spirit urges eternal labor and a redress of sin throughout the cosmos:

> Eternity is too short to finish the endless labor of a universe! I doubt if God himself has ever ended his work! From time beginningless to time eternal the march of labor goes on, from little sphere to larger, from lesser soul to greater, from duller intellect to one more brilliant! It is the changeless law for men and for angels alike! Onward, ceaselessly move onward, is the united aim of all dissevered instruments! There can come no pause. No man has ever moved an iota back since chaotic universes were spoken into moveless black of horrified space.[68]

Finally, the priest, the spirit, and Regan conclude with thunderous exclamation:

> "What is the limit of our power?" asked Father Renaudin.
>
> "There is no bound to what knowledge humanity may attain, save the clog of sin! Time, strength and study will overcome all other barriers!"
>
> "Monarch of Fate is man! Many yet shall chain the stars; shall drive the harnessed world!"[69]

Thus ends one of the forerunners of the American space opera, with an ode to the infinite capability of humanity to harness the power of the heavens for its own purposes. If the anonymous *Man Abroad* described the political and economic aspects of emerging space expansionist thought, it is Florence Carpenter Dieudonné's *Rondah* that described the unlimited spiritual ambition that accompanies and motivates space expansionism as well. If there can be no pause, and no backward movement for humanity, then for Dieudonné, as it would be later for Wells, the choice in the end is either all the universe or nothingness.

Works of spaceflight fiction would continue to proliferate near the end of the nineteenth century, with many of them focused on Mars, motivated in part by the "Mars mania" that took hold during the favorable opposition of the planet for telescopic observation in 1892 and the subsequent popular interest in the "canals" of the red planet. *The Man from Mars: His Morals, Politics and Religion* describes a visit to Earth from a member of an advanced Martian society by virtue of "spirit projection."[70] The book is notable principally for its representation of Martian civilization as a genuine possibility; for the utopian projections onto it; and for the reputation of its author, California State Senator William Simpson, the second American politician to author an interplanetary travel tale after George Tucker. It is also notable for making use of the same trope of a Martian visiting Earth that would be employed to great, popular effect sixty-nine years later in Robert A. Heinlein's 1961 classic, *Stranger in a Strange Land.*

In 1893, the first interplanetary romance was written: *Unveiling a Parallel: A Romance*, penned by two women from Cedar Rapids, Iowa, Alice Jones and Ella Merchant.[71] The story features two different, sometimes contrasting, Martian feminist utopias, and the book makes for interesting reading to this day. The socially provocative story features a male space traveler who journeys to Mars. He explores two different Martian societies while debating moral and social differences with Earth throughout, and in the process he falls in love with two different Martian women in sequence, each one arguably embodying two different feminist ideals.[72] The work provides little detail on the process of reaching the planet, but it does describe a feeling of ecstatic achievement and wonder when alighting on a new world, as well as Martian societies that, like those on Earth were beginning to do at the time, search the skies for potential visitors from other worlds:

> I will just say, that the time consumed in making the journey was incredibly brief. Having launched my aeroplane on the current of attraction which flows uninterruptedly between this world and that, traveling was as swift as thought. My impression is that my speed was constantly accelerated until I neared my journey's end, when the planet's pink envelope interposed its soft resistance to prevent a destructive landing. I settled down as gently as a dove alights, and the sensation was the most ecstatic I have ever experienced. When I could distinguish trees, flowers, green fields, streams of water, and people moving about in the streets of a beautiful city, it was as if some hitherto unsuspected chambers of my soul were flung open to let in new tides of feeling. My coming had been discovered. A college of astronomers in an observatory which stands on an elevation just outside the city, had their great telescope directed toward the Earth, just as our telescopes were directed to Mars at that time, and they saw me and made me out when I was yet a great way off.[73]

The unnamed male narrator, after flying his "aeroplane" to the red-pink planet—twenty years before the Wright Brothers' flight—initially finds life on Mars nearly identical to the nineteenth-century America that he has left behind, except with many of the gender roles reversed. He falls in love with Elodia, a successful banker and investor and sister to a leading astronomer, but he later discovers that Elodia freely indulges in drugs and alcohol and, along with other powerful Martian women, frequents a licensed pleasure palace called Cupid's Gardens to enjoy the services of the male prostitutes who work there. He loses his ardor for Elodia after discovering this and travels to another part of Mars, where he finds a society more to his liking, one where everyone collaborates, where women and men are truly equal, where material possessions are meaningless, and where he finds a new ideal companion, Ariadne, with whom he also falls in love before tearfully leaving for Earth. As the first explicitly feminist work of interplanetary utopia, *Unveiling a Parallel* remains one of the more interesting American spaceflight narratives of the nineteenth century.

Although these are perhaps the most interesting of the 1890s Mars narratives, there were others as well. Gustavus Pope's *Journey to Mars: The Wonderful World* featured Martians transporting the story's hero from Earth to their home world in an "ethervolt car," and it was a popular enough space adventure that it merited a sequel, *Journey to Venus: The Primeval World.*[74] A Nebraskan author and inventor named Henry Oleric, wrote another utopian Mars narrative, *A Cityless and Countryless World: An Outline of Practical Co-operative Individualism,* in which a Martian visitor arrives via an undescribed spacecraft to bring news of his utopia to Earth.[75] This work is notable for its optimistic description of the human potential for spaceflight: "Events that seem incredible, incomprehensible and impossible in one age, often become credible, comprehensible and possible in a succeeding age."[76] Olerich even implied that interplanetary travel was a relatively simple matter of utilizing contemporary knowledge and improving known technologies:

> If the earth and Mars, when in conjunction, were only a mile apart, a body could easily, even with your present knowledge of dynamics, be projected out of the reach of gravitation of one of these planets into that of the other. The actual interplanetary distance, which I traversed between the earth and Mars, calls in nothing new in kind, but only in degree. So you see that in order to be

able to make this interplanetary journey, you need only to improve on what you already have; and time has given the inhabitants of Mars, on this point, the advantage over the inhabitants of the earth.[77]

By the 1890s, then, the genuine possibility of spaceflight was emerging as a belief—at least as expressed in fiction—in multiple places across the country. Throughout the narratives covered here, it was predominantly through the efforts of individuals that the possibility of spaceflight was realized. The dominance of intrinsic, personal motivations in nineteenth-century American spaceflight narratives had reached the apogee of its solar system–wide macroeconomic scale in *Man Abroad*, and it had found grand spiritual expression in *Rondah*. However, it would take a bona fide American multimillionaire to present the most definitive vision of its potential microeconomic realization.

A Journey in Other Worlds, by John Jacob Astor IV, is a unique document in the history of American space exploration. For a start, it is the first description of spaceflight by an individual of immense personal wealth—the thirty-year-old John Jacob Astor IV, heir to the fur trade fortune of his great-grandfather. Astor's work—the only book by the American multimillionaire—represents the culmination of key trends in nineteenth-century American spaceflight fiction, including the emphasis on the private sector as the source of funding and motivation for spaceflight; the increasingly solar system–wide scope of spaceflight adventures; and the increasing social and cultural acceptance of spaceflight overall. A man deeply familiar with the world of power and money, Astor was also the first to explicitly present the realization of spaceflight as a public-private partnership, albeit one that is far more private than it is public. As such, followed only by Garrett Serviss's *Edison's Conquest of Mars*, Astor's novel marks the penultimate milestone in the nineteenth-century American conceptual development of spaceflight. It is a milestone that deserves to be more thoroughly considered and analyzed, as this vision of interplanetary travel in the twenty-first century—by the scion of the dominant high-society family of Gilded Age New York—contains some of era's boldest technological speculations as well as one of the most personal expressions of nineteenth-century American space expansionism.

John Jacob Astor IV was born into the loftiest heights of American wealth and status. His domineering mother, known to her contemporaries as "*the* Mrs. Astor," was the focal point of New York high society, while his father was a sophisticated but often absent and rakish man, who lavished his attention on sailing and yacht parties, including those on his steam yacht, one of the largest of the day, the *Nourmahal*—Arabic for "the light of the Harem."[78] John Jacob Astor IV's personal share of the Astor fortune was estimated at some $125,000,000 in 1889 (equivalent to $255 billion in 2024 GDP ratio terms, and $34.2 billion in 2024 PWC ratio terms), and the moneyed opulence that surrounded the young man was unrivaled.[79] When his imposing mother deigned to attend a Vanderbilt debutant ball in 1883, Mrs. Vanderbilt spent around $100,000 (equivalent to $3.2 million in 2024 CPI ratio terms) on the event to better impress her, while the marriage of the youngest of his four sisters, Caroline, is reported to have resulted in almost $1,000,000 in wedding gifts in 1884 (equivalent to $33 million in 2024 CPI ratio terms).[80] Astor himself was introduced to New York society at a lavish reception for eight hundred guests in 1887.

The young heir, however, had a greater passion for science and invention than he did for the debutant balls and high society outings that surrounded him. One acquaintance recalled that he was "too shy to talk well" and that the only thing he seemed to enjoy was working with his hands and stated that perhaps "if he had been poor he would have been happier, because he could have been a mechanic."[81] Throughout his life, his interests drifted toward the technical, spending hours on the mechanisms, models, and blueprints of his various inventions—which ranged from a Pneumatic Road Improver, capable of blowing dirt off roads, to his Vibratory Disintegrator, which was designed to obtain gas from peat moss.[82] Although he gave the rights to his invention patents to the public as a gift, he also maintained an active, though decidedly unsuccessful, commercial interest in technology, investing well over $250,000 (equivalent to $484 million in GDP ratio terms and $70 million in PWC ratio terms) in failed technology ventures, including the Nikola Tesla Company, the Amphere Electro Company, the Mutual Automatic Telephone Company, the International Power Storage Company, the Honduras Railroad Company, and the Keeley Motor Company.[83] At Astor's death the combined stock of all these companies was appraised to be worth less than the 265 copies of *A Journey in Other Worlds* that sprawled across the numerous libraries of the Astor residence, each of which was valued at $1.50.[84] There is some merit in this appraisal of value, as the far-reaching vision of his book arguably exceeded by a significant margin the intellectual merit of either his inventions or his technological investments.

The story of *A Journey in Other Worlds* centers on the actions of three main characters: President Bearwarden of the Terrestrial Axis Straightening Company; Dr. Courtland, a geologist, astronomer, and government-appointed examiner of the aforementioned company; and Richard Ayrault, the wealthy young vice president and principal stockholder of the company, who, like Astor when he wrote the book, is about thirty years of age. The Terrestrial Axis Straightening Company was formed for the purpose of eliminating Earth's tilt—and thereby eliminating the inconveniences of the seasons—by repeatedly moving large portions of ice from pole to pole in the counterdirection of the tilt until the planet's axis was perpetually perpendicular to the ecliptic. As the book opens, this great achievement is nearing its completion, and the principals are discussing what to do next now that most of the major problems of the world have been resolved, in a manner reminiscent of the men of the Baltimore Gun Club. They are discussing the idea of harnessing the newly discovered principle of *apergy*—a negative gravity concept that Astor had picked up from English author Percy Greg's *Across the Zodiac*—to circularize Earth's orbit around the Sun, when the young Ayrault has an epiphany and enthusiastically declares that they could instead use apergy to build a spacecraft and visit the planets.[85] All readily assent to the exciting plan and are soon off on the spaceship *Callisto* for myriad adventures, visiting a Jupiter overwhelmed with dinosaurs and a spirit-haunted Saturn—planets that serve as platforms for Astor's speculations about evolutionary biology and cosmic spirituality, respectively—while passing by and commenting on comets and Mars along the way, before returning to Earth with public fanfare and a celebration of Ayrault's marriage to a beautiful, scientifically inclined young Vassar student.

Although it may superficially sound like one, *A Journey in Other Worlds* was no idle playboy fancy. Astor's vision of the Earth in the year 2088 displays an abundantly creative and optimistic intellect, one that saw the development of spaceflight as the natural result of the exponentially increasing rate of technological progress that he observed around him. The young multimillionaire spent two years researching and writing the book—reportedly reading over one hundred books on science, astronomy, and spaceflight fiction in preparation—and undertook the work in the hope that it might "excite interest in scientific progress."[86] The result was a story weak in plot and character, but full of scientific facts, futuristic speculations, and an absorbing vision of the world in the twenty-first century. In the book, the astronomer-geologist Dr. Courtland describes his era as the most wonderful the world has as yet seen, and he notes how, with "the spread of republican ideas, and the great advance in science, education has become universal, for women as well as for men, and this is more than ever a mechanical age."[87] Electrical devices had replaced all animal and manual labor. High-speed electromagnetic railways and electric airplanes enabled rapid international travel, while people drove their pneumatic-wheeled, aluminum-framed "electric phaetons" on divided highways equipped with police speeding cameras.[88] There are perceptive descriptions of these technologies, as well as of those that enabled a network of videophones to encompass the globe, those that turned rainmaking into "an absolute science," and those that made the Sahara a global source of electrical power through the combination of steam boilers and giant concave mirrors.[89]

Dr. Courtland notes that the tremendous expenditures on armaments in previous centuries "were clouds with silver linings" that led to great improvements in metallurgy and machines. They also created such instruments of destruction as flying machines, high explosives, and gas bombs, which so horrified the human race that no further hostilities were ever contemplated; as a result, the long expected "Great War" never came.[90] In short, the "perfection of civilization" had been achieved, and there was wide hope that "the child of science with fadeless and growing splendour . . . may yet bring us and our little planet much closer to God."[91] For Astor, these developments were not extrapolated out of fancy. Rather, he projected them as being the result of what he saw as the exponential increase in technological capabilities: "Since every discovery or advance in knowledge increases our chance of obtaining more, it becomes cumulative, and our progress is in geometric instead of arithmetical ratio."[92] This observation of exponential growth as applied to the accumulation of knowledge and technology is one of earliest of its type. Along with the wide array of other technological speculations contained in *A Journey in*

Other Worlds, it shows that Astor had clearly spent significant time contemplating the future and was excited to hasten its arrival.

With his interests and intellect, it should be no surprise that Astor—who wrote *A Journey in Other Worlds* while ensconced in a new, $1.5 million family mansion with an estimated $750,000 in furnishings—would consider the expensive business of spaceflight a reasonable project for a private individual.[93] The impetus in the book comes from the young Ayrault, who, bored with the easy and abundant luxuries of Earth, and seemingly channeling Astor's own ennui, longs for something more: "This world will soon be a dull place. I wish we could leave it for a change. . . . Man has been a caterpillar for untold ages; can he not become the butterfly?"[94] This line of thought results in Ayrault's invention of the spacecraft *Callisto*, and he soon becomes convinced that the future of humanity lies in its development: "We are all tired of being stuck to this cosmical speck, with its monotonous ocean, leaden sky, and single moon. . . . Its possibilities are exhausted; and just as Greece became too small for the civilization of the Greeks, and as reproduction is growth beyond the individual, so it seems to me that the future glory of the human race lies in exploring at least the solar system."[95] Although other writers before Astor had considered these possibilities, there is a real feeling here of personal urgency and frustration; there is an almost palpable sense that, for the first time, someone desperately wanted their own spaceship. Although *A Journey in Other Worlds* is a fiction, it seems that this sentiment in particular is likely one that Astor himself would have voiced at a Gilded Age cocktail party. Indeed, in reading the book, one gets the sense that if one were to ask John Jacob Astor IV if he considered himself a space expansionist, he would have thought about it a moment, and then answered, "Indubitably."

Astor clearly thought of a spaceship as something that a wealthy man would build himself, but at the same time, perhaps because of his greater proximity to money and power than previous authors, Astor is also the first to suggest that the government might have a role to play, albeit a fairly small one. The first suggestion that the American government might be willing to finance the construction of a spaceship is a notion that Astor puts into the mouth of the secretary of the Navy—Secretary Deepwaters, who is also a participant in the scotch-and-water-infused discussion that produces the spaceship epiphany. The young Ayrault, however, dismisses the offer, and declares his desire to personally finance the exploration of space:

> "Do you know," said Deepwaters to Ayrault, while rapidly making his cocktail disappear, "the Callisto's cost with its outfit will be very great, especially if you use glucinum, which, though the ideal metal for the purpose, comes pretty high? I suggest that you apply to Congress for an appropriation. This experiment comes under the 'Promotion of Science Act,' and any bill for it would certainly pass."
>
> "No, indeed," replied Ayrault; "the Callisto trip will be a privilege and glory I would not miss, and building her will be a part of it. I shall put in everything conducive to success, but will come to the Government only for advice."[96]

Although Ayrault does indeed personally finance the construction contract for the twenty-five-foot wide, twenty-one-foot tall, two-story spacecraft (see figure 5.2), which the wealthy fellowship designs, the government is depicted as providing meaningful assistance. The secretary of state instructs all ambassadors and consuls to "telegraph the department anything they may know or learn that will be of use in adjusting the batteries, controlling the machine, or anything else, and will turn over to you in a succinct form all information that may be relevant."[97] The secretary of the Navy offers a farewell salute, rescue services in case they fail, and the promise to build numerous spaceships and open regular interplanetary travel lines once the group has demonstrated the feasibility of the endeavor and the habitability of Jupiter and Saturn. The president and the cabinet also give their moral support to the project, attending the launch-day festivities at Van Cortland Park, along with a million other Americans. To the mind of an American multimillionaire in 1894, a public-private partnership between the US government and wealthy American entrepreneurs seemed the obvious path for the nation to follow in its pursuit of the development of spaceflight capabilities in the late twenty-first century.

Astor's *A Journey in Other Worlds* thus represents a seemingly optimistic voice in favor of the view that the private sector is the proper wellspring of American spaceflight and space development. It is impossible to ignore, however, the fact that this same work also underlines some of the potential dangers of a space future fully under private control. Astor was evidently as moved by the prospect of colonizing the solar system as many of his predecessors and his successors—referring to it explicitly as

5.2 An American billionaire imagines *A Journey in Other Worlds* in 1894. *Source:* John Astor, *A Journey in Other Worlds* (New York, 1894).

"the manifest destiny of the race."[98] Ayrault dismisses Mars and Venus as potential prospects for colonization, with Mars being too small and with the axes of both being too highly inclined: "It would be as if colonists tried to settle Greenland, or had come to North America during its Glacial period."[99] Jupiter and Saturn, however, are thought by Ayrault to be another matter. Convinced of this, Dr. Courtland waxes poetic about the possibility of new expansion and colonization: "'As every place seems to have been settled from some other,' said Courtland, 'I do not see why, with increased scientific facilities, history should not repeat itself, and this be the point from which to colonize the solar system.'"[100] So far, so standard in terms of the expression of space expansionist ideas. A disturbing reminder of the dark side of colonization, however, follows shortly on the heels of this, with the secretary of the Navy stating that, as planetary colonization is only likely to occur sometime in the future, there will meanwhile be time to "absorb or run out all inferior races, so that we shall not make the mistake of extending the Tower of Babel."[101] The secretary of state promptly makes light of this statement from the secretary of the Navy—gibing that he is "putting on his war-paint" and that he "will soon want a planet to himself"—but beyond this, Astor does not seem to think that this casual reference to racial extermination merits further discussion. It is also a reminder that Astor himself was an active agent of American foreign conquest, having given an entire battery of artillery to the Army as a gift during the Spanish-American War, and having personally financed, drilled, and provided fitted uniforms from Wanamaker's for the Astor Light Battery before it fought in the Philippines.[102] Although Astor's Gilded Age racialism and amateur militarism were relatively common during his time and among his community, they nonetheless cast a dark shadow on *A Journey in Other Worlds*. The full context of *A Journey in Other Worlds* should therefore also serve as a reminder that the private development and operation of crewed spacecraft allows wealthy patrons to steer the course of space development toward their own beliefs and interests—for better and for worse.

John Jacob Astor IV's vision from 1894 nonetheless represents a milestone in the conceptual development of spaceflight. For the first time, an individual with the resources and inclination to invest in spaceflight expressed an eagerness to do. For the first time, the potential role of the government was recognized in the form of a public-private partnership. Most importantly, though, *A Journey in Other Worlds* gave the idea of spaceflight a level of social acceptance that it had not previously enjoyed. The very fact that an Astor—and the son of *the* Mrs. Astor, no less—had written such a book lent the idea credibility and even social cachet. Reviews outside of New York were more critical, but those from within the burgeoning commercial metropolis were positively enthusiastic, with one declaring that "large public attention" had been centered on the book for weeks.[103] Much of this enthusiasm was no doubt the result of the fact that the Astor family was the de facto landlord for many of the reviewers, but the cultural impact of the work was nonetheless real. Astor's well-known book thus helped lay the groundwork for the almost wholly positive public reception that liquid fuel rocketry pioneer Robert Goddard received when his work and ambitions became known to the world. Indeed, it is interesting to consider what type of relationship the two men might have developed had John Jacob Astor IV not decided, after helping his wife and others to the lifeboats, to go down with the *Titanic* in 1912 at the age of forty-eight, and what the combined forces of the Astors and the Guggenheims—who later became Robert Goddard's primary financial sponsors—together might have achieved for the development of American rocketry.

By the end of the nineteenth century, then, space expansionist thinking had already begun to enter the mainstream of American culture and had found its first—but far from its last—phenomenally wealthy evangelist. John Jacob Astor IV summed up what he saw as being the motivation for the pursuit of spaceflight in the twenty-first century:

> Notwithstanding our strides in material progress, we are not entirely content. As the requirements of the animal become fully supplied, we feel a need for something else. Some say this is like a child that cries for the moon, but others believe it the awakening and craving of our souls. The historian narrates but the signs of the times, and strives to efface himself; yet there is clearly a void, becoming yearly more apparent, which materialism cannot fill. . . .
>
> There is, so far as we can see, no limit to our journey.[104]

Interestingly, this expression of space expansionism does not refer to a need for new frontiers—though Frederick Jackson Turner had articulated his view of the importance of the supposed "closing" of the American frontier the year before. Rather, Astor describes it as being the product of a basic, limitless human yearning that remains even after all material wants have been satisfied. In this way, he argues that the movement into space will be driven by the intrinsic motivations of individuals who are following "the awakening and craving of our souls," in an early articulation of the view that spaceflight is part of ultimate human self-actualization.

Overall, the history of the nineteenth-century American conceptual development of spaceflight, and the expressions of space expansionist ideas and ideals that accompanied it, defies simple categorization and neat conclusions. There are stories of post–Civil War American social conflict extending to the Moon, a United States of America that encompasses the entirety of the solar system, feminist utopias on Mars, and narratives of middle-class journeys to the Moon and elite adventures to Jupiter and Saturn written by a leading social reformer and an heir to one of the largest fortunes in the country, respectively. Although there was great diversity and creativity in these tales, it should be acknowledged that there was only a limited sense of there being an intergenerational community of thinkers developing the concept of spaceflight at the time. With the notable exceptions of Poe's footnoted essay on earlier concepts and Astor's extensive research for his book, the concept of spaceflight was advanced by individual American virtuosos who were predominantly isolated from each other by time and space. Their works are more indicative of the potential of the individual spaceflight theorist at a given time and place than they are of an evolving American spaceflight community or conversation. That phase would await the emergence of the first rocket societies in the decades between the First and Second World Wars, as well as the emergence of pulp science fiction publications by Hugo Gernsback and others. In the first half of the nineteenth century, these early imaginative leaps above the atmosphere were like one-off suborbital flights: impressive, and demonstrative of significant talent, but not yet the technical basis for the development of the technologies and vehicles needed for the exploration of the solar system. By the end of the nineteenth century, however, spaceflight fiction had made an impression on the American zeitgeist, with new stories of voyages in space appearing every year in book form, in popular magazines, and serialized in newspapers.

The extent to which the major figures treated here were also rare intellectual talents is impressive. The conceptual development of spaceflight in the nineteenth century did not advance through the gradual addition of numerous, small contributions, but through the infrequent production of genuinely original works by brilliant minds. What sets the American experience apart in this sense is not so much the quality of these minds—which certainly found their equals in Jules Verne, Georges Le Faure, and Henry de Graffigny in France; Hugh MacColl, Percy Greg, and Robert Cromie in Great Britain; Hermann Ganswindt and Kurd Lasswitz in Germany; and Nikolai Federov and Konstantin Tsiolkovsky in Russia—but rather their quantity. It was the regular stream of significant American intellects, from Washington Irving to John Jacob Astor IV, attracted to the possibility of spaceflight that allowed it to develop to such a high level.

What then can we say in summary of these nineteenth-century American spaceflight narratives? First, the dominance of intrinsic motivations and private funding in the nineteenth-century American vision of spaceflight is striking. As NASA and US space policy in the twenty-first century move toward more integrated public-private partnership models for space exploration, it is important to recognize that this is not a new idea, but rather a return to the older American concept of space exploration expounded in the nineteenth century. The American history here is indicative of the broader history of the concept of spaceflight: Not once in almost three hundred years of fictional and nonfictional representations of spaceflight, from the seventeenth century to the late nineteenth, is there evidence that it was depicted as a governmental affair. It was always described as the product of the efforts of independent groups and individuals. It is not until the arrival of John Jacob Astor IV's *A Journey in Other Worlds* that the idea that the government might develop spaceflight is even hinted at, and even there Astor was rather confident that it would be wealthy individuals like himself that would first take up the challenge. Even Tsiolkovsky, writing in the early days of the Soviet command economy, identified a private group of wealthy individuals as the initial motivative force for his space migration in *Beyond the Earth*.[105] Prior to the twentieth century, the pursuit of spaceflight was seen as a personal endeavor—a fulfillment of intrinsic desires beyond the base material needs of mankind.

What then can we say about the evolution of space expansionism in this period? The latter half of nineteenth-century American spaceflight narratives include striking articulations of several of the core tenets of space expansionism, including the desirability of extending human society to all possible planets across the solar system and the idea of willful political experimentation as something that worlds beyond Earth can uniquely offer. With the period also introducing the idea of human social conflict extending to the Moon, many of the basic elements of American space expansionism have now been publicly articulated—not as a fully formed philosophy expressed by a single individual, but rather in aggregate through the works and writing of many different people, who nonetheless began to express similar ideas and ideals. In some respects, there is even an argument to be made that the late nineteenth century was a high-water mark for American space expansionist thought and enthusiasm. Whereas the anonymous writer of *Man Abroad* could, given the astronomical knowledge of the time, plausibly imagine human societies thriving on every planet in the solar system, we know now that in fact very few—if any—of these planets present particularly promising options for human settlement. Indeed, it is hard not to notice that the dramatic spaceflight visions of the late nineteenth century—with their stories of middle-class lunar vacations, and regular human adventures to Mars and Venus, and vast solar system–wide nations—are ones that we are still not particularly close to realizing even today, well over a century later. There is even an argument to be made that many of the grand plans of twentieth- and twenty-first-century space expansionists stem from the ideas and ambitions that were cultivated through the spaceflight narratives of this earlier era, but which the planetary probes and discoveries of the following century have made increasingly difficult to imagine turning into reality. Or, to paraphrase John Maynard Keynes's legendary aphorism: Practical men and women in the American space community, who believe themselves to be quite exempt from any intellectual influence, may still be the pawns of some defunct space expansionist.

Nonetheless, with the general concept of spaceflight firmly established—and communicated through stories that regularly placed the power to make it happen in the hands of a motivated and resourceful individual—the time had arrived for someone to begin to construct in reality what Americans had been imaginatively developing for a century, and what minds across the world had been developing for over three centuries. Four years after the publication of *A Journey in Other Worlds*, a brilliant teenager in Edward Everett Hale's old parish of Worcester, Massachusetts, would decide to dedicate his life to the goal of building a vehicle that could travel into space. This decision was specifically inspired by his reading of two tales of interplanetary travel, one British in origin and the other American. These two spaceflight narratives would be the final links in the intergenerational chain of influence and knowledge transfer that would result in the genesis of the physical development of the spaceship.

6

ALL THE UNIVERSE OR NOTHINGNESS

Dim and wonderful is the vision I have conjured up in my mind of life spreading slowly from this little seed-bed of the solar system through the inanimate vastness of sidereal space.

—H. G. Wells, *The War of the Worlds*, 1898[1]

Although there are many candidate moments for the origins of spaceflight rocketry, there is a good case to be made that a particularly crucial moment occurred in Worcester, Massachusetts, in 1898. At that place in space and time, a teenager named Robert Goddard read a series of spaceflight-related stories in *The Boston Post*, which, over the course of two months, helped inspire such a fascination with the spaceflight narrative and its possibilities that he dedicated his entire life to making it happen, later inventing and flying the world's first liquid fuel rockets as a direct result. This significant moment stemmed from two of the most original and consequential narratives about spaceflight: *The War of the Worlds* and *Edison's Conquest of Mars*, written by two of the most innovative spaceflight storytellers of their age, the pioneering British science fiction author H. G. Wells and the innovative American astronomy popularizer Garrett P. Serviss.

The convergence of the lives of these three early space pioneers—Wells, Serviss, and Goddard—represents the culmination of early spaceflight fiction and space expansionism, and a major transition point in spaceflight development. Prior to this moment, all the work that had contributed to the conceptual development of spaceflight had been done in the worlds of books and in the world of the mind. After this moment, however, while the spaceflight narrative continued to influence and shape space development, the actual development of rockets, satellites, and space vehicles begins in earnest. These stories therefore represent both a sort of endpoint and a starting point: the end of the first great era of the theoretical and conceptual development of spaceflight, and the beginning of the first great era of spaceflight's realization.

The War of the Worlds is the oldest story covered in this book that continues to be told anew, almost continuously, up until the present day. Whereas *From the Earth to the Moon* was last adapted into a film in 1958, the same year as the creation of NASA, *The War of the Worlds* has continued to be remade and retold for new audiences on a regular basis, with the most recent as of the time of writing being the *War of the Worlds* film adaptation from 2025.[2] As one of the most significant of the early spaceflight narratives, *The War of the Worlds* is worth examining in detail, along with the unparalleled life of its author, H. G. Wells, although there are so many elements in Wells's varied career that the primary focus here shall be on his biographical connections to science, astronomy, and spaceflight. In addition, because it was the unauthorized American version of *The War of the Worlds* that Robert Goddard read as a teenager in Worcester, Massachusetts, I will quote from that version of the story, with its action and scenes transposed to the Boston area, rather than the original, canonical version that takes place in and around London.

Wells was born in 1866, one year after the original publishing in France of *From the Earth to the Moon*, and he lived one of the most unusual, and unusually successful, literary lives of anyone in the nineteenth and early twentieth centuries. His mother was a housekeeper for the wealthy, most notably at the genteel estate of Uppark House, which he visited frequently during his teenage years, and his father was a gardener, outdoorsman, and occasional professional cricketer. It was from his father, who liked to lie outside at night looking up at the stars, that he learned an early love of sky watching.[3] It was from his mother's work at Uppark, however, that he acquired his first astronomical equipment, discovering an abandoned box there in the attic that contained the pieces of a telescope. At the age of fourteen, he assembled it, set it up in his room, and began to study the night sky.[4]

Wells severely broke his leg at age seven, and the months that he spent afterward convalescing instilled in him a deep love of reading, which, along with playing imaginary war games with himself, became his primary childhood hobby. The latter may have played a role in the visceral imagination that he applied to the horrors of a Martian invasion of Earth, while the former meant that he became well-educated quite early, winning a scholarship in 1884 to the prestigious Normal School of Science (later part of Imperial College London), where he had the great fortune to study biology under the legendary Thomas Huxley during Huxley's last year of teaching. Wells idolized Huxley, one of the foremost of the early advocates of the theory of evolution, and it is hard not to see him as a source of inspiration for Wells's later bold speculations on both far-future and extraterrestrial biology.

It was at the Normal School of Science that Wells acquired his significant scientific education and began to publicly speculate about the changes that might come about to both humanity and technology in the future. In his second year, he presented a paper entitled "The Past and Future of the Human Race," in which he described a distant future in which humanity would evolve into small-body creatures that were effectively hopping heads, that lived as "great unemotional intelligences," and that dispensed with regular food consumption and simply immersed themselves in tubs of nutritive fluid, all while living underground with artificial lighting to shelter from the cold on the surface of a cooling Earth.[5] When he later reworked and published the talk in 1894, Wells sounded a note that would become a refrain for many who would think about human futures in space in the century to come: "Undoubtedly it is a long time yet, but a long time is nothing in the face of eternity, and every man who thinks of these things must look eternity in the face."[6]

Although he was initially interested in science and biology—and evolutionary biology in particular—he was inconsistent in his science exam results, to the point where he failed his class in astronomical physics, and by his third year at the Normal School of Science he found himself more interested in history and the social sciences. After attending some meetings of the newly formed Fabian Society, and hearing talks by

George Bernard Shaw and Graham Wallas, he was also motivated to give a talk on the topic of democratic socialism to his college debating society. In particular, he emphasized the socialist arguments of Henry George, whose thinking had already infused the politics of the solar system in *Man Abroad* in the 1870s.

His time at the Normal School of Science was formative and gave him an intellectual foundation in science that he would employ in his works of speculative fiction. In December 1886, he set up *The Science Schools Journal* and was its first editor, and it was to that same journal, after leaving London to take up a teaching job in the country, that he would submit his installments of "The Chronic Argonauts," a time travel story that would eventually become *The Time Machine*. It was also to his old college debating society that he gave a talk titled "Are the Planets Habitable?" in 1888.[7] He also began to develop a public reputation when his essay "The Rediscovery of the Unique" was published in *The Fortnightly Review* in July 1891, and he received praise on it from no less than Oscar Wilde. Interestingly for a figure who would later become so closely tied with the idea of extraterrestrial life, the essay declared that "we have no reason at all to expect life beyond this planet," before ending with an extended soliloquy about the limited, and somewhat disillusioning, role that human science can play in the vast scale of the universe:

> Science is a match that man has just got alight. He thought he was in a room—in moments of devotion, a temple—and that his light would be reflected from and display walls inscribed with wonderful secrets and pillars carved with philosophical systems wrought into harmony. It is a curious sensation, now that the preliminary flutter is over and the flame burns up clear, to see his hands lit and just a glimpse of himself and the patch he stands on visible, and around him, in place of all that human comfort and beauty he anticipated—darkness still.[8]

The implied tension between the light of science and a darkness enveloping humanity is one that would infuse Wells's works throughout his life.

The publication of "The Time Machine" in 1895 in *The New Review* catapulted Wells almost instantaneously into becoming a household name and a popular writer. The work sold over ten thousand copies in book form and was rapidly translated into French. Although there had been time travel stories before, none had been affected with as much depth of thought on the far-future logical consequences of evolution—including biological, planetary, and stellar evolution. The book is famous for its pioneering depiction of a time travel machine and for its depiction of humanity evolving into two separate species—Eloi and Morlock—which represented evolutionary adaptations of the gentrified and working classes of Wells's time. From the perspective of the development of the spaceflight narrative, however, the most significant innovation is the extensive role that far-future sky watching plays in the book, both to mark the passage of time and to provide opportunities for reflection on the vastness of the cosmos and the continuous process of change within it. The time traveler, for example, looks at the stars above him in the year 802,701 AD, and finds them simultaneously alien and comforting:

> Above me shone the stars, for the night was very clear. I felt a certain sense of friendly comfort in their twinkling. All the old constellations had gone from the sky, however: that slow movement which is imperceptible in a hundred human lifetimes, had long since rearranged them in unfamiliar groupings. But the Milky Way, it seemed to me, was still the same tattered streamer of star-dust as of yore. Southward (as I judged it) was a very bright red star that was new to me; it was even more splendid than our own green Sirius. And amid all these scintillating points of light one bright planet shone kindly and steadily like the face of an old friend. Looking at these stars suddenly dwarfed my own troubles and all the gravities of terrestrial life. I thought of their unfathomable distance, and the slow inevitable drift of their movements out of the unknown past into the unknown future. I thought of the great precessional cycle that the pole of the earth describes. Only forty times had that silent revolution occurred during all the years that I had traversed. And during these few revolutions all the activity, all the traditions, the complex organizations, the nations, languages, literatures, aspirations, even the mere memory of Man as I knew him, had been swept out of existence.[9]

Here, eighty-five years before Carl Sagan's "Pale Blue Dot" speech in *Cosmos*, is a similarly evocative, if slightly more melancholic, precursor to that same cosmic sentiment.

The Time Machine also includes some of the first popular speculation about the future of humanity, or lack thereof, in the context of stellar evolution. As the time traveler proceeds ever farther into the future—more than thirty million years into the future—the sky he describes becomes ever more alien. The passage is so remarkable in historical context that it bears reading in full:

> As I drove on, a peculiar change crept over the appearance of things. . . . The alternations of night and day grew slower and slower, and so did the passage

> of the sun across the sky, until they seemed to stretch through centuries. At last a steady twilight brooded over the earth, a twilight only broken now and then when a comet glared across the darkling sky. The band of light that had indicated the sun had long since disappeared; for the sun had ceased to set—it simply rose and fell in the west, and grew ever broader and more red. At last, some time before I stopped, the sun, red and very large, halted motionless upon the horizon, a vast dome glowing with a dull heat, and now and then suffering a momentary extinction. . . . The earth had come to rest with one face to the sun, even as in our own time the moon faces the earth. . . . So I travelled, stopping ever and again, in great strides of a thousand years or more, drawn on by the mystery of the earth's fate, watching with a strange fascination the sun grow larger and duller in the westward sky, and the life of the old earth ebb away. At last, more than thirty million years hence, the huge red-hot dome of the sun had come to obscure nearly a tenth part of the darkling heavens.[10]

It is startling to recognize that Wells seems to be describing a future in which our Sun has evolved into a cooler, "duller" red giant. This is somewhat the inverse of what we expect to happen today; we expect the Sun to grow larger as the millennia pass but to increase its luminosity and total radiation output as it does so. Nonetheless, the result is ultimately the same: the eventual extinction of life on Earth due to the natural process of stellar evolution. Wells is admittedly off by two orders of magnitude in terms of the expected timeline as today the Sun is not expected to become a red giant for another five billion years or so. His description is nonetheless astounding as he wrote it almost thirty years before Arthur Eddington scientifically derived this fate for our home star in his book *The Internal Constitution of the Stars*, published in 1926.[11] How then did Wells anticipate this cosmic future so fully and so far ahead of his time? The key is likely the teacher for the astronomical physics class that he failed. That teacher was none other than Norman Lockyer, the leading theorist of stellar evolution of his day and a pioneer of stellar spectroscopy. Lockyer had discovered helium in the Sun's spectrum in 1868 and had contributed seminal work on stellar evolution in the 1870s, before taking up a teaching post for astronomical physics at the Normal School of Sciences in 1885. Perhaps Wells's failure in Lockyer's class was due to too much creative thought as Lockyer never specifically wrote that the Sun would become a red giant. Wells was thus significantly ahead of his time, not only in foreseeing this future but in writing a best-selling popular book about it and thus informing the public about this likely fate for our Earth long before the scientific establishment would.

Within a year of the publication of *The Time Machine*, Wells was also working on *The War of the Worlds*. He had moved with his second wife, Amy Catherine "Jane" Robbins, to the town of Woking, and there in a small house he began to write out the story by hand. The origins of *The War of the Worlds* came from a suggestion from Wells's brother, Frank. He suggested the idea for the story because he knew about Wells's interest in life on Mars and had been reflecting on the then-popular myth of the extinction of the Tasmanian Aboriginal peoples.[12] Wells dedicated the book, "To My Brother Frank Wells, This Rendering of His Idea."[13] The idea from his brother seems to have combined with an idea Wells expressed in *The Time Traveler*, in which the time traveler muses about the possibility that, at some point between his time and the distant future, humankind might evolve into "something inhuman, unsympathetic, and overwhelmingly powerful."[14] Wells began to imagine that these types of beings might come not from humanity's future but rather from another planet, and that they might treat humans as Europeans had treated others around the world, much as Washington Irving had also suggested at the beginning of the nineteenth century.

The War of the Worlds would first be serially published in *Pearson's Magazine* from April to December of 1897. Shortly thereafter, starting on January 10, 1898, and running daily until February 3, *The Boston Post* began to run an unauthorized, heavily edited serial version of the story that had the narrative transplanted to Boston and its surrounding towns. There are many differences between Wells original version and the version that was printed in *The Boston Post* and read by the young Robert Goddard. Even the famous first line of the story, "No one would have believed in the last years of the nineteenth century that human affairs were being watched keenly and closely by intelligences greater than man's and yet as mortal as his own," was changed, ending instead with "as immortal as his own" in the American version—perhaps a change the newspaper felt was needed for a more religiously minded New England readership.[15] Nonetheless, the story is the same, with almost all the same story beats and key descriptions; the main exception was the locations described. In the original, the first Martian cylinder lands at Woking, due to H. G. Wells living there when he wrote the story; in the *Boston Post* version, the first cylinder lands in Concord, Massachusetts, and the path of destruction goes through Lexington and Bunker Hill, in a direct homage to the battles, and route taken by the British, at the start of the American Revolution.

In *The War of the Worlds,* Wells suggested that humanity had thus far demonstrated a languid complacency when faced with the vastness of the cosmos. In the first installment of the story, Wells declares: "No one gave a thought to the older worlds of space, or thought of them only to dismiss the idea of life upon them as impossible or improbable. At most, terrestrial men fancied there might be other men upon Mars—probably inferior to themselves, and ready to welcome a missionary enterprise."[16] This later sentiment in particular had already been seen across many earlier spaceflight narratives that included extraterrestrial beings, including the interest in sending "bibles to the moon" that had attended the Great Moon Hoax in the 1830s.[17] Wells, however, was eager to provoke people into thinking about the far-reaching consequences of the new theories of both stellar and biological evolution. Indeed, the motivation for the Martians in *The War of the Worlds* is arguably the first "survival of the species" argument for space travel in literature. In a section that was removed in the *Boston Post* version, Wells writes that "since Mars is older than our earth, with scarcely a quarter of the superficial area, and remoter from the sun, it necessarily follows that it is not only more distant from life's beginning but nearer its end. The secular cooling that must some day overtake our own planet has already gone far indeed with our neighbour," and, therefore, "to carry warfare sunward is their only escape from the destruction that generation by generation creeps upon them."[18] This is arguably the first existential survival motivation for spaceflight ever expressed. Wells describes the Martians as having understood their ultimate fate and then correspondingly prepared for an invasion of a more habitable world, noting that "had our instruments only permitted it we might have seen the gathering trouble far back in the 19th century."[19] With such an existential motivation, Wells also suggests that the Martians were willing and able to engineer a planetary-scale coordinated effort to save themselves, much as he would later argue that humanity would ultimately need to do as well, when its own international, technological, and planetary challenges would necessitate the emergence of a global world government, or what Wells would call the *world state*.

Although Wells does not go into detail on the spaceflight launcher technology of the Martians, he does provide an unprecedented description of what terrestrial astronomers might see of a large-scale launch system on another planet. He appropriately recognizes that *Mars opposition*—when the Earth is directly between Mars and the Sun—marks the best time for an expedition between the two worlds:

> As Mars approached opposition . . . Lavelle of Java set the wires of the astronomical exchange palpitating with the amazing intelligence of a huge outbreak of incandescent gas upon the planet. It had occurred toward midnight of the twelfth, and the spectroscope, to which he had at once resorted, indicated a mass of flaming gas, chiefly hydrogen, moving at an enormous velocity toward this earth. . . . He compared it to a colossal puff of flame, suddenly and violently squirted out of the planet, "as flaming gas rushes out of a gun."[20]

The unnamed narrator of the story learns of this development when he runs into a local astronomer, Ogilvy, and describes how the astronomers then observe the same flame emanating from Mars each night for ten days. We can only wonder about the massive planetary-scale launch system that Wells envisioned. For the *Boston Post* serialization, these story elements were helpfully provided in summary descriptions that ran before each installment, and they give a remarkably matter-of-fact account of the imagined interplanetary invasion: "As Mars approached opposition to this world and was consequently only 40,000,000 miles from this earth instead of 140,000,000 miles, great masses of fire were shot from the Planet directly toward us. They moved upon the earth with enormous velocity. Many men of science called attention to these shots—ten in number. One of these Things, resembling cylinders of tremendous size, struck in Concord, Mass, just over the Lincoln line."[21]

The War of the Worlds is the first full story of the invasion of Earth by beings from another planet, the short speculations of Washington Irving notwithstanding. In the story, the narrator describes watching as the first Martian transportation cylinder makes its way toward Earth in a moment of peace before the invasion fundamentally shifts humanity's relationship with the external cosmos: "One night, the Thing then could scarcely have been 10,000 miles away. I went for a walk with my wife. It was starlight, and I explained the signs of the Zodiac to her, and pointed out the bright dot of light creeping zenithward, toward which so many telescopes were pointed. For, in those days, there was no terror for men among the stars."[22] In many ways, the vast array of space horror and alien invasion narratives—from *Alien* to *The Dark Forest*—begin here. At first, the response to the landing of the Martian cylinder is one of cosmic kinship, with the astronomer Ogilvy trying to help the Martians, who he fears may be in trouble: "With a quick mental leap he linked the thing with the flash upon Mars! The thought of the confined creature was so dreadful to him that he forgot the heat and went forward to the cylinder to help [open it]."[23] Characteristically, in the *Boston Post* edition, the local astronomer Ogilvy is the only member of the astronomical community mentioned, in a classically leveling American edit, whereas in Wells's original there is also reference to more elevated astronomers and astronomical institutions, including the Astronomer Royal, the Astronomical Exchange, and even the Lick Observatory. The response in the story to the discovery of the cylinders is initially optimistic, with press reports of "a message received from Mars" and an expectation that, as had been anticipated by most prior spaceflight fiction, any Martian life would be more or less human: "I think everyone expected to see a man emerge—possibly something a little unlike us terrestrial men, but in all essentials a man. I know I did."[24] Instead, Wells introduces into the spaceflight narrative the first themes and images of cosmic existential horror that would be developed in the decades to come more fully by H. P. Lovecraft and H. R. Geiger: "I presently saw something stirring within the shadow, grayish billowy movements, one above another, and then two luminous discs like eyes, . . . The body heaved and pulsated convulsively. A lank tentacular appendage gripped the edge of the cylinder; another swayed in the air. Even at this first encounter, this first glimpse, I was overcome with disgust and dread."[25]

What follows from that moment is a description of abject destruction. Dozens are killed by the Martians "swiftly and silently" with the aid of a heat ray and a parabolic mirror. Although the narrator initially believes that the Martians will be able to be overcome due to the difference in gravity between Mars and Earth, the mechanical devices that the Martians have prepared for their invasion more than compensate for their gravitational disadvantage. After the Martians ignore a white flag of surrender,

6.1 Graphics from *The War of the Worlds* (top) and *Edison's Conquest of Mars* (bottom). *Source: The Boston Post*, 1898.

the narrator retreats—in the *Boston Post* version, via Walden Pond—and then encounters the invaders in their full and now infamous form (see figure 6.1): "A monstrous tripod higher than many houses, striding over the young trees and smashing them aside in its wallowing career; a walking engine of glittering metal, reeling now across the grass."[26] This image is now so familiar to us that it takes some work to imagine how imaginative and provocative it would have been in the late 1890s, when Queen Victoria still reigned in England and William Mckinley was serving as the twenty-fifth president of the United States. Amid it all, Wells continues to remind the reader of the broader cosmic context by including acts of sky watching in the narrative: "The storm had left the sky clear, and over the smoke of the burning land the little fading pin-point of Mars was dipping into the west."[27]

The narrative proceeds with a survey of the destruction wrought. The destruction of Concord, Lexington, and Waltham is complete—with seemingly all the notable buildings in each town getting its own mention in *The Boston Post.* The survey is so complete, in fact, that it is hard not to wonder if there might be an early example of product placement in the newspaper when it comes to the destruction of Waltham Watch Company and the melting of the "hundreds of thousands of little pieces of brass and nickel and steel used in the manufacture of the 400,000 watches which are usually in the course of construction at the same time in the factory."[28] The Martians then proceed to unleash a mysterious black powder that has deadly effects beyond those of their heat ray, and they destroy an ironclad and a steamer in the Boston harbor, before finally visiting destruction upon Boston itself. Yet Wells does not only focus on destruction; he makes it clear that the Martians are builders as well. There is not one type of machine that the Martians inhabit and operate, but many. There is the iconic tripod fighter machine, but also a flying machine (see figure 6.2), a handling machine, and an excavating machine, with it noted that "these last had brought with them certain fresh appliances that now stood in an orderly manner about the cylinder."[29] He even depicts the Martians as establishing a sort of brick-production facility, with the excavator machine digging up clay and making bars of it, in perhaps the first concrete example of the application of in situ resource utilization (ISRU) technology in fiction. Wells does not describe Martians as a species interested in wanton destruction for its own sake, or as beings who are in some way "evil," but rather more plainly as a conquering and constructive species who are simply moving sunward to rebuild their civilization as their own planet reaches the end of its life-carrying capacity.

After fifteen days, the story draws to its conclusion, with its famous ending that features the fearsome and conquering Martians felled by simple terrestrial bacteria. The first signs of this come as the narrator crosses Harvard Bridge, which is almost buried in the dying remains of an alien red weed that the Martians had brought with them and that had covered large swaths of the landscape. As the narrator proceeds into the ruins of

6.2 Images of Martian flying machines in *The War of the Worlds. Source: The Boston Post* (left) and *Pearson's* (right), 1898.

Boston, he comes across a dozen Martians lying dead in a row, "slain by the putrefactive and disease bacteria, against which their systems were unprepared; slain, after all man's devices had failed, by the humblest things that God has put upon this Earth."[30] Although the story of the Martian invasion of Earth is one of the most fascinating in all of spaceflight fiction, in some respects it is the literal and figurative postmortem to the story that is more interesting for what it reveals about Wells's views about spaceflight and space expansionism.

After the invasion, Wells describes the scientists of the world as investigating and learning from what the Martians have left behind. They conclude that "spectrum analysis of the black powder points unmistakably to the presence of an unknown element with a brilliant group of three lines in the green, and it is possible that it combines with argon in forming a compound which acts at once with deadly effect upon some constituent in the blood."[31] Extraterrestrial biology thus is added to the human store of knowledge and even to museums, with the general public later becoming "familiar with the magnificent and almost complete specimen in spirits at the Museum of Natural History, and the countless drawings that have been made from it."[32] With the death of the Martians, they become objects of scientific interest, rather than fear, once again.

Perhaps the most striking passage from the perspective of the development of the spaceflight narrative, however, comes at the very end of the story. Wells suggests that Earth should remain concerned about potential future invasions from Mars, and he suggests that the Martians may already have traveled to and conquered Venus—and that if they have not, then perhaps humans should consider conquering it themselves. Demonstrating once again Wells's strong understanding of the orbital dynamics involved, the narrator notes that "at present the planet Mars is in conjunction, but with every return to opposition, I, for one, anticipate a renewal of their adventure."[33] He also suggests a sort of planetary defense observation program: "It seems to me that it should be possible to define the position of the gun from which the shots are discharged; to keep a sustained watch upon this part of the planet, and to anticipate the arrival of the next attack."[34] He then proceeds to discuss the evidence that suggests that the Martian had landed on Venus, noting specifically that scientists

> had advanced excellent reasons for supposing that the Martians have actually succeeded in effecting a landing on the planet Venus. Seven months ago now Venus and Mars were in alignment with the sun—that is to say, Mars was in opposition from the point of view of an observer on Venus. Subsequently a peculiarly luminous and sinuous marking appeared on the unilluminated half of the inner planet, and almost simultaneously a faint dark mark of a similar character was detected upon a photograph of the Martian disk. One needs to see the drawings of these appearances in order to appreciate fully their remarkable resemblance in character.[35]

Wells is not only the first to describe a spaceflight expedition from Mars to Venus, but also the first to imagine how an interplanetary expeditionary force might signal back to its home planet via planetary-scale signaling.

The last words of *The War of the Worlds* that the young Robert Goddard would have read in *The Boston Post* would have been provocative ones for a young mind in 1898. The fifteen-year-old would have read, after processing all the horror and wonder that had come before: "At any rate, whether we expect another invasion or not, our views of the human future must be greatly modified by these events. We have learned now that we cannot regard this planet as being fenced in as a secure abiding place for man; we can never anticipate the unseen, good or evil, that may come upon us suddenly out of space."[36] Goddard would also come to conclude that this planet was not a secure abiding place for man. He would go on to pen his own text about the far future, *The Last Migration*, in which he discussed the need for humanity to someday travel to other stars in order to preserve itself. In Wells's original text, which Goddard would have read later in his life, he continued with more fulsome, space expansionist prose:

It may be that in the larger design of the universe this invasion from Mars is not without its ultimate benefit for men; it has robbed us of that serene confidence in the future which is the most fruitful source of decadence, the gifts to human science it has brought are enormous, and it has done much to promote the conception of the commonweal of mankind. It may be that across the immensity of space the Martians have watched the fate of these pioneers of theirs and learned their lesson, and that on the planet Venus they have found a securer settlement.[37]

From here, Wells, showing his beginnings as not just a spaceflight storyteller but arguably the first true philosopher of space expansionism, lays out the existential argument for spaceflight:

The broadening of men's views that has resulted can scarcely be exaggerated. Before the cylinder fell there was a general persuasion that through all the deep of space no life existed beyond the petty surface of our minute sphere. Now we see further. If the Martians can reach Venus, there is no reason to suppose that the thing is impossible for men, and when the slow cooling of the sun makes, as it must do at last, this earth uninhabitable, it may be that the thread of life that has begun here will have streamed out and caught our sister planet within its toils. Should we conquer?[38]

It is hard to overemphasize the significance of this statement within the intellectual history of spaceflight and space expansionism. Never before had any writer connected the idea of spaceflight with the inevitable fate of the Earth to become ultimately uninhabitable due to the future stellar evolution of our Sun. Here, in 1898, Wells not only makes that connection but also suggests that, like the Martians in the story, we too may someday need to seek out new homes on other planets. Although there are many arguments and rationales for space expansionism, and many of them had been expressed in earlier eras, the intellectual history of the existential argument—that we must expand to the stars or humanity will ultimately go extinct—begins here.

Wells not only recognized that interplanetary travel was technically possible but also asked the fundamental question that is posed by this potentiality: What should humanity do about it? He finishes *The War of the Worlds* with a recognition that a future in space may belong to others if it does not to humanity: "Dim and wonderful is the vision I have conjured up in my mind of life spreading slowly from this little seed bed of the solar system throughout the inanimate vastness of sideral space. But that is a remote dream. It may be, on the other hand, that the destruction of the Martians is only a reprieve. To them and not to us perhaps is the future ordained."[39] Wells thus concludes his tale with perhaps one of the most enduring images of space expansionism: the idea of humanity as a seed from earth that can be planted across not only the rest of the solar system but also across the rest of the universe. He would go on to write with even greater clarity about this vision of space expansionism throughout his life. The impact of his story of Martian invasion would also be significant and personal.

After helping to inspire Robert Goddard to become the inventor of the first liquid fuel rocket, Wells would live long enough to watch liquid fuel German V-2 rockets rain down around him in London during the Second World War—rockets that had been developed in part with the aid of Goddard's publicly available patents. However, before that would all come to pass, someone else was called to answer his question of "Should we conquer?" Three days after the conclusion of the serialization of *The War of the Worlds* in *The Boston Post*, in a heavily promoted edition of *The Boston Sunday Post* (see figure 6.3), a new serialized story began immediately that would answer that question decidedly in the affirmative: *Edison's Conquest of Mars*.

The author of that work, Garrett Putnam Serviss, was not a global household name like H. G. Wells, but he did have some local fame in the United States as one of the great American astronomy popularizers of the nineteenth century. He was born in 1851 in Sharon Springs, New York, and like Wells, he benefited from the early acquisition of a telescope—in this case, as a gift from his older brother. He obtained a science degree from Cornell University in its inaugural class. As a Cornell student, he demonstrated a flair for poetry and history, and as a senior he delivered a commencement day speech titled "The Perpetuity of the Heroic Element."[40] Serviss

6.3 Advertisements for *The War of the Worlds* and *Edison's Conquest of Mars*. Source: *The Boston Post*, 1898.

then went to Columbia University to study law and was admitted to the New York State bar in 1874. Shortly thereafter, however, he decided instead to pursue journalism as a career, eventually becoming the night editor of the New York newspaper *The Sun*. It was in the pages of *The Sun*—where Serviss got to moonlight for years as the paper's popular and anonymous astronomy columnist—that he developed his full passion for conveying to the public the discoveries and wonders of astronomy. In 1892, at age forty-one, he decided to resign from his editorship and dedicate himself full-time to the popularization of astronomy. It was a productive career transition. Clyde Fisher, the first director of New York City's Hayden Planetarium, would later declare that Serviss had done "more to popularize astronomy than any one in America, and perhaps in the entire world."[41] Although writing was the primary source of his influence, his first project upon dedicating himself to astronomy was perhaps his most innovative and groundbreaking: an original stage production entitled *The Urania Lectures* that established him as a new type of figure in astronomy—the popular astronomical entertainer.

Serviss's *Urania Lectures* was a groundbreaking sound-and-light spectacle that demonstrated that the combination of astronomical knowledge, spaceflight narrative, and visually-stunning cosmic vistas could form a source of popular entertainment. *The Urania Lectures* had resulted from an offer to Serviss by the wealthy industrialist Andrew Carnegie to fund a popular traveling theater show based on his successful astronomy columns. Carnegie would later go on to finance George Ellery Hale and the Mt. Wilson and Mt. Palomar Observatories, but it was his investment in *The Urania Lectures* that was his first philanthropic patronage for astronomy. When Serviss left *The Sun*, this project was his immediate focus, and through it he developed America's first large-scale magic lantern performance about space. In Carnegie's new Music Hall in New York in 1892, magic lantern projections, transparent painted scenery, and the new wonder of electric stage lightning all combined to transport audiences into space in way never seen before. Audiences were treated to visuals of solar eclipses—as seen from both the Earth and the Moon—as well as dramatically lit lunar surfaces as part of Serviss's showcase production, *A Trip to the Moon*.[42] *A Trip to the Moon* was not the first theater production of a lunar voyage. That honor goes to Jacques Offenbach's operetta *Le Voyage dans la lune* in 1875. However, Serviss's performance was one of the first major theatrical events that put the image of cosmic vistas from a nonterrestrial perspective front and center. It was, in this way, a precursor to Georges Méliès's *A Trip to the Moon* from 1902, as well as Frederick Thompson's immersive theme park attraction "Trip to the Moon" at the Pan-American Exposition in Buffalo in 1901 and at Coney Island in 1902.[43]

Serviss took his audience within twenty-four thousand miles of the Moon through models and stage-lighting, and he explicitly anchored the audience within the long history of the spaceflight narrative:

> If we could travel to the moon, as Hans Pfaall did in Poe's story with the aid of a balloon, or borne by the wings of a flock of wild swans after the manner of Domingo Gonzales' lunar journey, or darting through space inside a hollow projectile according to Jules Verne's method, we should find many things to interest us along the way . . . At the distance of 24,000 miles from the moon's centre we should be able, without any optical aid, to perceive the true character of its surface. Already the great craters, or mountains, which give so strange an appearance to the moon would be clearly discernible. . . . We could then choose our point of landing upon its surface and continue our journey, being henceforth under the gravitative dominion of the moon.[44]

The origins of Serviss's *A Trip to the Moon* actually began in Germany in 1889 at the Urania Institute in Berlin, where William Foerter, a student of Alexander von Humboldt, and Max Wilhelm Meyer had created stage sets and scenes for a production called *From the Earth to the Moon*.[45] These sets were acquired by the Music Hall and formed the basis of an even grander production, with the addition of lecture material and narration from Serviss. The New York production ran three times a week, with tickets available for one dollar (equivalent to $35.20 in 2024 CPI ratio terms) and an estimated cost of production of $20,000 (equivalent to $5.5 million in 2024 PWC ratio terms). After running in New York, *A Trip to the Moon* and Serviss then toured America, traveling from city to city, with the production and its scenes making it at least as far as Denver.[46] Imagining Serviss traveling across America with his German-imported trip to the Moon sound-and-light show in the 1890s gives us a glimmer of perhaps one of the greatest space showman stories never fully told.

When Arthur Brisbane, editor of *The New York Evening Journal*, decided that he was going to acquire the rights to publish H. G. Wells's *The War of the Worlds* for his newspaper, he decided that he could capitalize on the opportunity by also commissioning and publishing a sequel to the story—and so he turned to the leading astronomical popularizer in the country, Garrett Serviss. Brisbane, ever the canny promoter with an eye for an angle, encouraged Serviss to call up Edison in hopes that they would collaborate on the sequel. Although Edison decided not to collaborate, he agreed, ever the canny promoter himself, to let Serviss and Brisbane publish a story that featured him as a brilliant interplanetary hero. *Edison's Conquest of Mars* thus was born. The narrative ran daily simultaneously in *The New York Evening Journal* and in *The Boston Post*. The story was heavily publicized in advance, with *The Boston Post* in particular taking liberties with its emphasis on the "collaboration" between Edison and Serviss (see figure 6.3). With or without Edison's collaboration, however, the story that unfolded from Serviss's mind and pen was an astoundingly original encapsulation of the spaceflight narrative, one that encompassed many of the elements that had come before and that also introduced a number of new ideas, ones that both space mission planners and science fiction writers draw on to this day. Although less famous and culturally impactful than *The War of the Worlds*, *Edison's Conquest of Mars* in fact included an even greater number of technical innovations related to the conceptual development of spaceflight.

The narrative begins in many respects where *The War of the Worlds* leaves off, with a destroyed earth, and with scientists learning from the technologies left behind by the Martians. In Serviss's story, not all of the Martians perished; a "few survivors fled in one of their projectile cars" by means of a mysterious explosive force that gave their vehicle seven miles per second of velocity in order to overcome the attraction of the Earth and the resistance of the atmosphere.[47] Right from the start, Serviss is educating the reader on the basic physics of spaceflight, which may well have given the young Robert Goddard some of his earliest technical information on the subject. After the Martian invasion, the mood across the Earth is one of despair, and it becomes even more despairing still when it is announced that astronomers have observed lights on Mars indicating that they are preparing for another invasion. In response, in what is essentially the first assemblage of an *Avengers*-like cast for a space adventure, "a few dauntless men of science . . . Lord Kelvin, the great English savant; Herr Roentgen, discoverer of the famous X-ray, and especially Thomas A. Edison, the American genius of science" team up to examine "the Martian engines of war" and learn their secrets in order to perfect their own "ships of space" and, just as importantly, their own disintegrator rays.[48] Edison perfects the flying machine (figure 6.4) and in it undertakes "an experimental trip to the moon," which Serviss describes briefly. At the same time, he introduces one of the more enduring spaceflight metaphors—the ocean of space: "By visiting this little nearby island in the ocean of space, Mr. Edison simply wished to demonstrate the

6.4 Images of Edison's flying machine and Earth's spaceship fleet in *Edison's Conquest of Mars. Source: The Boston Post*, 1898.

practicability of his invention, and to convince, first of all, himself and his scientific friends that it was possible for men—mortal men—to quit and to revisit the earth at their will."[49]

Serviss's core spaceflight technology is a type of antigravity device that is derived from the electric repulsion of atoms, a not-uncommon type of imagined solution to the problem of spaceflight as we have seen. He recognizes that a spacecraft would ultimately have to be highly controllable, noting that as they approach the Moon they would "manipulate the electric charge upon our car" in such a way as to counter the Moon's gravity "in order that we might gradually approach it and with an easy motion, settle, without shock, upon its surface."[50] With this new spaceflight technology now being perfected by Edison and his scientific avengers, the people of Earth decide to strike back at the Martians and "take possession of that detestable planet."[51] As the narrator points out, this had as much, if not more, to do with the development of the disintegrator than the spacecraft; "this enthusiasm would have had but little justification had Mr. Edison done nothing more than invent a machine which could navigate the atmosphere and the regions of interplanetary space."[52] Already by the late 1890s, then, the idea of the spacecraft had become so common that it is no longer presented as sufficient for enthusiasm by itself. Now, for the first time, the spacecraft is presented first and foremost as a tool of war, complete with new fantastical weaponry, and it is this new application of the spacecraft that is to become a new source of enthusiasm.

With the spaceship and disintegrator developed by Edison, the global desire to launch a counterattack on Mars spurs a historic meeting in Washington, DC, of all the world's leaders, who convene to discuss that all-important prerequisite for any serious space project—funding: "The salvation of the planet, as everyone was now convinced, depended upon the successful negotiation of a gigantic war fund, in comparison with which all the expenditures in all of the wars that had been waged by the nations for 2,000 years would be insignificant. The electrical ships and the vibration engines must be constructed by scores and thousands."[53] In an early recognition of the true economic scale that would be required to realize spaceflight, Serviss presents an individual—Thomas Edison—as being able to achieve lunar flight with a prototype, but notes that "to multiply these upon a war scale was not only beyond the resources of any individual—hardly a nation on the globe in the period of its greatest prosperity could have undertaken such a work. All the nations, then, must now conjoin."[54] This thesis, that global unity would be required to realize the spaceflight narrative on its grandest scales, is here explicitly introduced for the first time. H. G. Wells had obliquely implied it in *The War of the Worlds* from the perspective of the Martians, and he would go on to develop it more fully later in his life in his writings on the world state, including in *The World Set Free* in 1913.[55] In 1898, however, it was from Serviss, not Wells, that readers would connect the idea of global governance and spaceflight, and it would be Washington DC, rather than London, that would be presented as the new capital of the planet, as it grew into a "gigantic city" to accommodate the "great congress of nations," where the kings, queens, emperors, presidents, and prime ministers of the countries of the world would strategize and plan for Earth's response to the attack from Mars.

At the congress of the nations, the unnamed US president gives a speech that encourages the governments of the world to fund a large space program with which to invade Mars:

> "We have come together," he said, "to consider a question that equally interests the whole earth. I need not remind you that unexpectedly and without provocation on our part the people—the monsters, I should rather say—of Mars, recently came down upon the earth, attacked us in our homes and spread desolation around them. Having the advantage of ages of evolution, which for us are yet in the future, they brought with them engines of death and destruction against which we found it impossible to contend. . . . But we have also another object. Whether we send a fleet of interplanetary ships to invade Mars or whether we simply confine our attention to works of defense, in either case it will be necessary to raise a very large sum of money. None of us has yet recovered from the effects of the recent invasion . . . It will be part of our business here to raise a gigantic war fund by the aid of which we can construct the equipment and machinery that we shall require. This, I think, is all I need to say. Let us proceed to business."[56]

The president suggests that $10 billion dollars would be needed for the purpose of this interplanetary war fund, with a Pacific Coast Senator suggesting afterward that $25 billion would be better. It is not exactly clear in the story how much the congress of nations ultimately raises, but either number is a staggering sum within the context of the era and the spaceflight narratives of the nineteenth century. The former amount is comparable to $15.7 trillion dollars today and the latter comparable to $39.2 trillion

Table 6.1 Interplanetary war fund by nation in *Edison's Conquest of Mars*

Nation	Value of contributions
United States	$2,000,000,000
Great Britain	$1,500,000,000
Germany	$1,000,000,000
All Others	$5,500,000,000 to $21,500,000,000
Total	$10,000,000,000 to $25,000,000,000
2024 US PWC ratio equivalent	$2,850,000,000,000 to $8,480,000,000,000
2024 US GDP ratio equivalent	$15,700,000,000,000 to $39,200,000,000,000

Source: Garrett P. Serviss, *Edison's Conquest of Mars* (Carcosa House, 1947).

dollars today in relative 2024 GDP ratio equivalent terms. This is an unprecedented amount of suggested funding for spaceflight, which today would equal around 14 to 35 percent of the total GDP of the Earth for a year.[57] Needless to say, humanity has yet to realize this suggested level of resource allocation for spaceflight. The funding would also go entirely to manufacturing and production costs for the new fleet of spacecraft, as in the story Edison had already demonstrated the prototype spacecraft and disintegrator ray (see figure 6.4). This funding is therefore not for spaceflight per se, but rather for the production of spacecraft that can wage interplanetary war. Although the funding provided is part of a united planetary effort, there is still geopolitical signaling involved, with the Germans contributing $1 billion, the British not willing to be outdone and subscribing $1.5 billion, and the Americans leading the way with $2 billion (see table 6.1). Here again, as with Verne's work, a massive new space project is depicted as being funded by voluntary contributions—except that this time it is the heads of state and whole nations that are volunteering their funds rather than individuals. Once the sum is raised, Edison simply declares, "Give me *carte blanche* . . . and I believe I can have a hundred electric ships and three thousand disintegrators ready within six months."[58] The tradition of entrepreneurs arriving in Washington, DC, and declaring incredible space achievements to be possible if only they are given carte blanche and large amounts of public funding is therefore, at least in fiction, well over a century old.

Two thousand crew are allocated to the one hundred ships, and there is already the sense of the crew being a select group. The narrator noted that though they were few in numbers, "they represented the flower of the earth, the culmination of the genius of the planet," in language perhaps even more effusive than that used to describe the original Mercury Seven astronauts. Although the expedition has war as its primary purpose, Serviss makes clear that the expedition is both managed by, and in many respects in the service of, scientists and technologists: "The expedition had, of course, profoundly stirred the interest of the scientific world, and representatives of every branch of science, from all the civilized nations, urged their claims to places in the ships."[59] In particular, the leaders of the expedition on the flagship with Edison included Kelvin, Rayleigh, Roentgen, and Moissan, and "each of these men cherished hopes of wonderful discoveries, along his line of investigation, to be made in Mars."[60] In this way, *Edison's Conquest of Mars* is the first spaceflight narrative of a mission to Mars conducted not only by a crew of scientists and engineers, but by an international crew as well. Serviss evidently took as his model a similar scientific-military expedition from the then-not-so-distant past: "On the model of the celebrated corps of literary and scientific men which Napoleon carried with him in his invasion of Egypt, Mr. Edison selected a company of the foremost astronomers, archaeologists, anthropologists, botanists, bacteriologists, chemists, physicists, mathematicians, mechanics, meteorologists and experts in mining, metallurgy and every other branch of practical science, as well as artists and photographers."[61] Here then we are beginning to approach a truly modern sense of an interplanetary expedition, which—minus the military aim of the mission—would stand as the ideal for international spaceflight throughout much of the twentieth century.

Although the antigravity space launch technology that Serviss used in *Edison's Conquest of Mars* did nothing to advance the technical concept of spaceflight, he did introduce something truly novel and technically important: the space suit. Although Washington Gladden had suggested the need for respirators on the Moon, as amazing as it may seem, no prior work had described the unique need for a pressurized suit to do work in the vacuum

6.5 First illustration of space suits in *Edison's Conquest of Mars. Source: The Boston Post,* 1898.

of space (see figure 6.5). However, Serviss saw its need clearly, noting that Edison had solved the issue of the vacuum of space "by inventing an airtight dress constructed somewhat after the manner of a diver's suit, but of much lighter material. Each ship was provided with several of these suits, by wearing which one could venture outside the ship even when it was beyond the atmosphere of the earth."[62] Serviss also described how the people in the suits could communicate using telephonic mouthpieces and receivers in their suits and that they just needed to "connect themselves by the wires, and conversation could then be easily carried on."[63] If this seems a cumbersome way to imagine communication between space suits, it helps to remember that Serviss is writing this less than one year after Guglielmo Marconi's first radio transmission across the Bristol Channel in May 1897.

As the squadron departs Earth, they are treated to a massive, planetary-scale fireworks display, spreading out for hundreds of miles, of such magnitude that it "must certainly have been visible to the inhabitants of Mars."[64] Before they can proceed on their journey, however, fireworks of a different sort impede their progress; a meteor strikes one of the electric space cars at a speed of eighteen miles per second, destroying it and killing its crew. With this, in another innovation, Serviss introduces a topic that would await genuine space mission planning to address fully: the question of burial in space for crewmembers of an expedition. A decision is made, after accounting for their losses, that "instead of committing them to the fearful deep of space, where they would have wandered forever, or else have fallen like meteors upon the earth, to give them internment in the lunar soil."[65] On to the Moon the expedition then proceeds—not for discovery, but for ritual and honorable burial.

Serviss's Moon is simultaneously the desolate, airless world that was known to science at the time and a world of ancient fantasy. The squadron approaches the Bay of Rainbows and the narrator expresses dreadful wonder: "We had long been aware that the Earth's satellite was a body which had passed beyond the stage of life, if indeed it had ever been a life supporting globe; but none of us were prepared for the terrible spectacle which now smote our eyes." The Moon appeared to the crew as a "dead and mangled world."[66] Nonetheless, there was room for fantasy still in the 1890s, and in their explorations the crew uncovers "the ruins of what appeared to be an ancient watch tower. It was evidently composed of cyclopean blocks larger than any I had ever seen among the ruins of Greece, Egypt and Asia Minor."[67] The crew stays on the Moon for a couple of days to repair the wrecked ship and bury their dead. Before leaving, however, they discover a faint remnant of one of the ancient lunar inhabitants, a five-foot-long footprint, which gives rise to arguably the first scene of scientific investigation on the lunar surface in literature: "Around this gigantic footprint gather most of the scientific members of the expedition, wearing their oddly shaped air-tight suits, connected with telephonic wires, and the spectacle, but for the impressiveness of the discovery, would have been laughable in the extreme. Bending over the mark in the rock, nodding their heads together, pointing with their awkwardly accoutred arms, they looked like an assemblage of antediluvian monsters collected around their prey."[68] The descriptive similarity to some of the first scenes of lunar geology during the Apollo program is striking. Similarly, the still-hoped-for promise of lunar resources also found a place in Serviss's narrative in the form of giant crystals that would be "an inexhaustible future mine of wealth."[69] The expedition then leaves the Moon after only forty-eight hours—around twice as long as the crew of Apollo 11 spent on the lunar surface.

As the fleet continues on toward Mars, with meteors and the Moon behind them, they continue their astronomical tour of the inner solar system, starting with a comet and then landing on an asteroid. The comet—described as "lots of separate meteors, and blazing metallic particles and hot mineral vapours"—captures the electric-powered fleet in its own electrical field and carries them back to Earth, almost destroying them (see figure 6.6).[70] After a short twenty-four hours back on Earth, they are off again to Mars, expecting to arrive there in forty-two days at ten miles a second. Thirty-five days into their journey, however, they notice an asteroid that is inhabited by a small group of Martians. Serviss's Martians are very different than Wells's, being essentially giant humans with exaggerated features. They have the same destructive weaponry, however, and as the fleet approaches the asteroid, the Martian disintegrator ray destroys two of the electric cars. The fleet regroups, and prepares to attack: "The signal was given, and the circle of electric ships closed in upon the asteroid."[71] Such a phrase would not be out of place in pulp science fiction stories up through the 1950s and 1960s, and yet here it is in 1898—a reminder that we have been imagining, and consuming stories about, space fleet warfare since the age of Queen Victoria.

As part of the attack on the asteroid, Edison must leave the ship to get a better shot for his disintegrator; in the process, he exits through literature's first space airlock: "In the meantime, Mr. Edison had been donning his air-tight suit. Before we could clearly comprehend his intention he has passed through the double trapped door which gave access to the exterior of the car without permitting the loss of air, and was standing upon what served as the deck of the ship."[72] Once on the asteroid, the crew captures a Martian alive. They also discover—instructively, by weighing themselves and realizing that the density of the asteroid is much higher than they expected—that the asteroid is in fact made of pure gold, leading the president of an

6.6 Passengers observe a comet from a spaceship in *Edison's Conquest of Mars*. *Source: Boston Post*, 1898.

Australian mining company aboard the fleet to describe the asteroid as the "great gold mine of Mars."[73] This asteroid encounter also allows Serviss to show off his sophisticated understanding of orbital dynamics. He notes that the Martians would have a "harvest time" for the asteroid that would be determined by its closest approach to the planet, and in the fight with the Martians he describes how various crew members and Martians are sent on long elliptical orbits around the asteroid due to its low gravity, in another first: the first low-gravity fight in spaceflight fiction.

As the electric spaceship fleet approaches Mars, a new grand vision of the red planet is revealed. Serviss describes snow and water ice around the South Pole, which astronomers at the time believed they were seeing when they observed the polar caps of frozen carbon dioxide, and, of course, Serviss notes how "crossing and recrossing the red continent, in every direction, were the canals of Schiaparelli."[74] The Mars of Serviss, as was popularly expected at the time, is a world of massive canal engineering, which he described as "a double system, intended both for irrigation and for protection, and far more marvelous in its completeness than the boldest speculative minds among our astronomers had ever dared to imagine."[75] Mars is also depicted as red throughout, not because it is a desert, but because the trees, grass, and flowers are red as well. The Martians live in entirely metallic buildings and have their capital around the Lake of the Sun, a circular body of water five hundred miles in diameter with dozens of great canals running away from it like the spokes of a wheel. Serviss's Mars is both similar in some details to the Mars we know—he accurately points out that the Martian day is forty minutes longer than an earth day, for example—and yet also completely different and vibrantly lifelike. *Edison's Conquest of Mars* thus serves as a vivid reminder that when the original space exploration vision caught hold in the generation of Goddard, Oberth, and Tsiolkovsky, they expected to find inhabited and readily habitable planets next door, whereas the truth has turned out to be far different.

The human battle against the Martians on Mars is as dramatic as any that has been depicted since—from John Carter to John Carmack—and it includes a number of technical innovations. The Martians deploy a great smoke barrier to shield themselves from attack, but what is most fascinating is that one of the scientists in the fleet then collects some of the smoke into a jar so that chemists on board can examine it spectroscopically—a decidedly *Star Trek*–style example of the practice of science featuring directly in a space battle.[76] As the battle intensifies, the *Star Trek* similarities continue, with Edison deciding that their strategy should be to amplify their disintegrators and then attack where the density of the Martians is highest. Despite the great smoke barrier, one of the astronomers, who knows Mars like the back of his hand, is able to determine the location of the Lake of the Sun even through the smoke. The attack, however, is countered violently by the Martians, with the loss of thirty of the ships in the fleet and six hundred crew. It is at this point that Colonel Alonzo Jefferson Smith enters the narrative.[77] Smith suggests a plan to distract the Martians by drawing fire with the fleet, while a single ship sneaks around the back and attacks from behind. The narrator and Smith go on this mission and are the only crew to actually land on Mars in the expedition: "We believed we were the first of the descendants of Adam to set foot on the planet of Mars. . . . A thrill, born half of wonder, half of incredulity, ran through me at the touch of the soil of Mars. Here was I, actually on that planet, which had seemed so far away, so inaccessible, and so full of mysteries when viewed from the earth. And yet, surrounding me, were things—gigantic, it is true—but still resembling and recalling the familiar sights of my own world."[78] In their sneak attack, the crew happen upon a Martian storehouse, where they are shocked to see a human woman—Aina—singing for the Martians. They rescue Aina, and through a long description of the process of learning her language, and of the various attempts by the men of the crew to woo her, it is revealed that her ancestors had long ago been captured by the Martians during their time on Earth, when the Martians had also built the pyramids and the Sphinx, with the latter being revealed as a portrait of the Martian emperor.

While the crew takes the time to learn Aina's language, they hide from Martian eyes behind the Martian moon Deimos, in an early description of space battle positioning and tactics:

> Our present distance from the surface of Mars did not exceed 12,500 miles, and we had reason to believe the Martians possessed telescopes powerful enough to enable them not merely to see the electrical ships at such a distance, but to also catch sight of us individually. Accordingly strict orders were given, not only that the ships should be moored on that side of the satellite which is perpetually turned away from Mars, but that, without orders, no one should venture around on the other side of the little globe or even on the edge of it, where he might be seen in profile against the sky.[79]

In addition, in order to further shroud them from detection, we are also given the first description of space camouflage: "The sentinels all wore garments corresponding in color to that of the soil of the asteroid, which was of a dark, reddish brown hue."[80] The late nineteenth century thus appears to have included at least one thinker who imagined that thankfully still largely unrealized landscape of future space combat that has been so fully and enthusiastically elaborated by writers, filmmakers, and computer game programmers in the subsequent centuries.

The final attack on Mars can be fairly described as cinematic, despite having been written only two years after the first cinemas were built in 1896. The Martian's capital around the Lake of the Sun is targeted, and in a *Return of the Jedi*–style scene, the goal is to sneak into the Martian power station, aided by Aina, and turn off the power so that the gates that control the flow of the Martian canals are opened and a deluge of biblical proportions can overwhelm and destroy the cities and capital of Mars. This is achieved, with the help of an electrical expert from Edison's shop, and they watch from orbit as the flood envelops Mars:

"In a little while we had overtaken the first wave, which was still devouring everything. We saw it bursting the banks of the canal, sweeping away forests of gigantic trees, and swallowing cities and villages, leaving nothing but a broad expanse of swirling and eddying waters, which, in consequences of the prevailing red hue of the vegetation and the soil, looked, as shuddering we gazed down upon it, like an ocean of blood flecked with foam and steaming with the escaping life of the planet from whose veins it gushed."[81]

In the flood, they see swept away a giant, forty-foot woman—identified by Aina as an enslaved woman from Ceres, part of another alien giant species with whom the Martians were at regular war. After the flood, the crew lands at the Martian palace to secure the surrender of the malevolent Martian emperor, who presides over the military government of Mars. The emperor surrenders, but they end up disintegrating him anyway after they see him attack a woman in rage. With this, the conquest of Mars, by Edison and his scientific avengers, is complete.

The expedition had left the Earth with one hundred ships and two thousand crew, and they return from their triumph with but fifty-five ships and 1,085 souls. Despite the losses, there is little melancholy in the return, and mostly celebration: "When at length we once more saw our native planet, with its well-remembered features of land and sea, rolling beneath our eyes, the feeling of joy that came over us transcended all powers of expression."[82] As the ships approach Earth, they first come down near Mt. Fuji in Japan and proceed to drop off crew members in their home countries, making stops in Japan, China, Russia, Germany, Italy, and France. Finally, the fleet approaches the city of New York, which, in another remarkably prescient description—given that the city would become the future home of the United Nations—Serviss describes as "the new capitol of the world."[83] With this, one of the most unusual, bloody, and far-seeing spaceflight epics of the nineteenth century is brought to conclusion.

It was these two spaceflight narratives—*The War of the Worlds* and *Edison's Conquest of Mars*—that so enthralled Robert Goddard and played an important role in his decision to dedicate himself to the pursuit of spaceflight technology. Goddard's regular diary entries at the time of their publication do not make note of his reading the stories, occupied as he was with describing other daily adventures and trials related to buying, selling, and trying to melt aluminum sheets for his precocious, youthful aluminum balloon project. However, in his "Material for an Autobiography" that he wrote in 1933, he gives pride of place in his journey to the two stories:

> An event happened about this time which was destined to provide me with all the scientific speculative material that I could desire. In January of 1898 there appeared for several months, in the *Boston Post,* the story Fighters from Mars, or the World of the Worlds, in and near Boston. This, and also the story which followed it, Garrett P Serviss's Edison's Conquest of Mars, gripped my imagination tremendously. Wells's wonderfully true psychology made the thing very vivid, and possible ways and means of accomplishing the physical marvels set forth kept me busy thinking.[84]

There, in one of the clearest cases imaginable, we have the pioneering developer of liquid fuel rocketry crediting Wells and Serviss both for taking hold of his imagination and, just as importantly, inspiring him to think about how to accomplish the technological and physical feats described in their spaceflight tales. With this, the long line of spaceflight narratives—from the 1630s to the 1890s—that had served to advance the conceptual development of spaceflight from generation to generation had now transferred the accumulated knowledge and inspiration to an individual who would lay the engineering and experimental foundations for the realization of spaceflight.

We also know that Goddard maintained his interest in and connection to Wells and Serviss, as he wrote letters to them both and continued to read their works for the rest of his life. His diary entry for October 14, 1903, when he was twenty-one, shows that he wrote a letter addressed to "Prof. Garrett P Serviss," although it is unclear if he ever received a response.[85]

From his diary, we know that he read "Other Worlds" and "The Moon" by Serviss in 1913, and also read Wells's *In the Days of the Comet* and *The Shape of Things to Come* in 1927 and 1934. We also know that he kept a newspaper clipping of a Serviss article that must have meant a lot to him, published as it was a few years after Goddard went public with his paper titled "A Method for Reaching Extreme Altitudes." The article highlights the feasibility of exactly the type of rotating, solid fuel rocket launch to orbit that Goddard was working on: "There is another way in which a trip might be made from the earth to another planet, or from another planet to the earth, and that is by means of a repeating rocket. A rocket does not progress by pushing against something, but by the reaction of its exploding charge—just as a gun kicks. If it could carry charges enough it could keep on to the moon."[86]

By far the most touching note that Goddard made with regard to the influence of these stories on his life and career, however, is the letter that he wrote H. G. Wells in 1932. In it, Goddard tells Wells of his achievements and the role that Wells played in helping to inspire him, and it deserves to be quoted here in full:

In 1898 I read your *War of the Worlds*. I was sixteen years old, and the new viewpoints of scientific applications, as well as the compelling realism of the thing, made a deep impression. The spell was complete about a year afterward, and I decided that what might conservatively be called "high-altitude research" was the most fascinating problem in existence. The spell did not break, and I took up physics as a profession, entering the physics department of Clark University, Worcester, Massachusetts, in 1914, where I have been head of the department since 1923. My work on the theory of the space problem culminated in 1919 with a paper in the Smithsonian Miscellaneous Collections, and since that time many books on the subject have appeared, and four "Interplanetary Societies" have been organized, in as many different countries. Two years ago, my own work became more than a spare-time hobby, for through the enthusiasm of Colonel Lindbergh, a fund was provided by the late Daniel Guggenheim of New York City, to develop the high-altitude rocket for meteorological investigations. I transferred the work to New Mexico, where the large tracts of level unoccupied land, clear air, and equable climate provide more suitable conditions than did New England. How many more years I shall be able to work on the problem I do not know; I hope, as long as I live. There can be no thought of finishing, for "aiming at the stars" both literally and figuratively, is a problem to occupy generations, so that no matter how much progress one makes, there is always the thrill of just beginning. I have the greatest admiration for your later work, which you no doubt feel is more important than your writings of the nineties. What I find most inspiring is your optimism. It is the best antidote I know for the feeling of depression that comes at times when one contemplates the remarkable capacity for bungling of both man and nature.[87]

After such an incredibly personal and poetic letter, it is hard not to be a little disappointed by Wells's short response on May 3, 1932: "Thanks for your friendly letter. It's the sort of greeting we appreciate from people like you."[88] Nonetheless, this exchange reflects perhaps the earliest personal connection between a writer of spaceflight narratives and a builder of launch vehicles, both of whom were working on encouraging humanity to reach for the stars in their own respective ways.

Both Serviss and Wells would continue to write about space and spaceflight-related matters for the rest of their lives. Indeed, both of them would live long lives, Serviss until age seventy-eight, dying in 1929, and Wells until age seventy-nine, dying in 1946, surviving even Robert Goddard, who died in 1945. Serviss lived long enough to see some of Goddard's ideas for spaceflight rocketry become a popular topic of discussion, and Wells lived long enough to witness V-2 rockets shattering the windows of his Hanover Terrace apartment in London during the Second World War. Each of their lives had influence on the other's in ways large and small, and in ways directedly connected to spaceflight.

Serviss would continue to write popular articles about astronomy, as well as a number of science fiction books. His story *The Moon Metal* explored the idea of a new metal being discovered on the Moon—artemisium—which could replace gold on the Earth, with corresponding dramas of the personal and political-economy varieties.[89] Perhaps his most interesting contribution to the spaceflight narrative, however, beyond *Edison's Conquest of Mars*, was his novel *A Columbus of Space*, first published in 1909. The story revolves around an eccentric inventor who harnesses the power of "inter-atomic energy," which he uses to propel a spaceship that carries him and his friends on a journey to the planet Venus in arguably the first nuclear-powered spaceship in fiction.[90] The novel overall was dedicated to the readers of Jules Verne: "Not because the author flatters himself that he can walk in the Footsteps of that Immortal Dreamer, but because, like Jules Verne, he believes that the World of Imagination is as legitimate a Domain of the Human Mind as the World of Fact."[91] Through his wide-ranging works,

Serviss can be said to have had almost as much of an influence as Verne in helping to legitimate the world of imagination as a source of innovation in the development of spaceflight.

Wells would also continue to engage with space and the spaceflight narrative through the rest of his life as well, arguably in even more profound ways than in *The War of the Worlds*. Around the same time as *The War of the Worlds*, Wells also wrote two short stories: "The Crystal Egg," which described a sort of crystal ball that could see activities on the planet Mars; and "The Star," which describes the impact on the Earth of a trans-Neptunian object that causes the remnants of humanity to rebuild civilization near the poles.[92] Wells also wrote his own narrative of a lunar voyage, *The First Men in the Moon*, first published serially in 1900 and 1901. In the story, a novelist, Mr. Bedford, becomes acquainted with an eccentric physicist, Mr. Cavor, who has discovered an antigravity material that he terms *cavorite*. The principle of antigravity flight is essentially identical to that suggested by John Leonard Riddell in 1847: Working by the opening of various panels, the antigravity cavorite repels the sphere containing it in the opposite direction. This allows Mr. Bedford and Mr. Cavor to voyage to the lunar surface in a pressurized, windowed sphere, with compressed provisions and an apparatus to absorb their carbon dioxide.[93] With Wells already established as a writer on par with Jules Verne at the time of its publication, *The First Men in the Moon* was popular and influential, with its lunar surface adventures amid the Selenites—the insectoid, underground civilization inhabitants of the Moon—serving as the other major source of inspiration, besides Verne, for George Méliès's pioneering film *Le Voyage dans la lune* in 1902.

There is an argument to be made that Wells's most broadly important space-related work, aside from those he wrote directly related to spaceflight, was actually his *In the Days of the Comet*, published in 1906. Largely forgotten today, the book caused a major scandal for Wells due to the public perception that it was advocating for what at the time was called *free love*. In the book, the green vapors from a passing comet mix with the atmosphere of the Earth and cause dramatic changes in human affairs, include the creation of a *world state*.[94] Interestingly, shortly before Wells was born, during the passing of the Great Comet of 1861, evidence suggests that for two days of its orbit, Earth was actually within the comet's tail in a manner not dissimilar from the scenario in the book. Whether or not Wells explicitly connected the event with this fascinating moment in astronomical history is unknown, but Wells could have easily learned about this event through his education at the Normal School of Science. *In the Days of the Comet* features a jealous lover who plans to murder his romantic rival—but his entire psychology and worldview is transformed by the vapors from the comet, as is the case for everyone on the planet. The population of the world awakes after the comet's passing to find they now have a permanently greater degree of empathy and love, with the effect that the murderously jealous lover now finds that he can happily accept a more open, eventually even polyamorous, approach to his life and loves. Wells was highly critiqued for this story, and he fell out of favor with the high-minded Fabian Society over it. Although the Fabians were self-proclaimed socialists, they were also concerned that Wells was actively encouraging the concept of free love, which had been popularized by the Oneida Community, John Humphrey Noyes, and Victoria Woodhall in the 1850s to 1870s, and which the Fabians considered to be a dangerous idea. Although it is clear that Wells bore some significant social costs for the work, the narrative and its message were true to Wells's own unconventional lifestyle, and it is interesting that he portrayed the shift that leads to worldwide free love as being caused by something that comes from space. It is impossible to know the relationship between Wells's terrestrially derived cosmic perspective and Wells's uncommon approach to life and love, but it is interesting nonetheless to see one of the foremost of the early space philosophers begin to engage with the more personal implications of the social, interpersonal, and psychological changes that space may bring to humanity. Indeed, in associating space and the cosmos with radically different social arrangements and transformations in personal psychology and perspective, H. G. Wells seemed to anticipate—over a half century before Robert A. Heinlein's *The Moon Is a Harsh Mistress*, the Earthrise photo from Apollo 8, and Frank White's *The Overview Effect*—the even more significant role that spaceflight fiction and spaceflight itself would have on social and cultural dynamics in the twentieth and twenty-first centuries.

The most important works of space expansionist thought in Wells's life, however, were still yet to come. In *The World Set Free: A Story of Mankind*, from 1913, Wells expressed a clear view that he believed that humanity would and should extend its reach to new worlds in the cosmos. The book is most known for the prescient predictions it contained regarding the potential for destructive nuclear weapons, as well as the corresponding need, in Wells's view, to establish a world state to control and eradicate these terrifying new weapons while still harnessing nuclear power and bringing prosperity to the world in the process. However, it also contains one of Wells's clearest expressions of his belief in the principles of space expansionism:

> The air no longer imprisons us, this round planet is no longer chained to us like the ball of a galley slave . . . In a little while men who will know how to bear the strange gravitations, the altered pressures, the attenuated, unfamiliar gases, and all the fearful strangeness of space, will be venturing out from this earth. This ball will be no longer enough for us; our spirit will reach out . . . Can not you see how that little argosy will go glittering up into the sky, twinkling and glittering smaller and smaller, until the blue swallows it up. They may succeed out there; they may perish, but other men will follow them. . . . It is as if a great window opened.[95]

We should consider Wells to be not only a writer of spaceflight narratives but an early spaceflight philosopher as there seems to be a genuine worldview—and, in particular, a space expansionist worldview—behind much of his writing and his work. For Verne, for example, the trip to the Moon is predominantly a fun adventure, even if there are many lines in his work that invoke the grandeur of the cosmos and that indicate that further exciting adventures await those who can manage to travel into space. For Hale, even, life in space is a way to develop new ways of living together, but that is the focus, not an expansion of humanity beyond Earth. For Wells, however, it seems that he was in fact convinced that whether there was to be a human future among the stars was a vitally important question, and further that it was a question that needed to be answered positively.

In addition to his space expansionist musings at the end of *The War of the Worlds* and *The World Set Free*, Wells also advocated for a space expansionist worldview in the last paragraph of his best-selling nonfiction work *The Outline of History*. In the final chapter, entitled "Unification of the World," after laying out his eight "fundamentals of the coming world state," which remain worth reading for anyone interested in planetary affairs today, he ends on a romantic and expansionist note:

History is and must always be no more than an account of beginnings. We can venture to prophesy that the next chapter to be written will tell, though perhaps with long interludes of setback and disaster, of the final achievement of world-wide political and social unity. But when that is attained, it will mean no resting stage, nor even a breathing stage, before the development of a new struggle and of new and vaster efforts. Men will unify only to intensify the search for knowledge and power, and live as ever for new occasions . . . Gathered together at last under the leadership of man, the student-teacher of the universe, unified, disciplined, armed with the secret powers of the atom and with knowledge as yet beyond dreaming, Life, for ever dying to be born afresh, for ever young and eager, will presently stand upon this earth as upon a footstool, and stretch out its realm amidst the stars.[96]

If there remains any doubt, however, that Wells was a true space philosopher and a genuine space expansionist, one has but to watch the last scenes of the epic 1936 film *Things to Come*, based on Wells's book *The Shape of Things to Come.*

In the movie, after the "dictatorship of the air" has brought peace to the world following multiple tragic wars, the leadership of an underground city is preparing to launch humans around the Moon for the first time. At the same time, a rebellious group of citizens argues that this should stop, that the onrush of technological change is too much to allow for a peaceful life, and that they will stop this process by halting the first flight to the Moon. The final scene, after the son and daughter of two of the leaders of the city—Cabal and Passworthy—have launched on their journey to the Moon, is one of the most evocative and declarative passages on the choice before humanity as it ponders a future among the stars:

CABAL: There! There they go! That faint gleam of light.

PASSWORTHY: I feel that what we've done is monstrous!

CABAL: What they've done is magnificent.

PASSWORTHY: Will they come back?

CABAL: Yes, and go again and again, until the landing is made and the moon is conquered. This is only a beginning.

PASSWORTHY: But if they don't come back, my son and your daughter, what of that Cabal?!

CABAL: Then presently others will go.

PASSWORTHY: Oh god, is there ever to be any age of happiness? Is there never to be any rest?

CABAL: Rest enough for the individual man, too much and too soon, and we call it death . . . but for man, no rest and no ending. He must go on, conquest beyond conquest. First this little planet in its winsome ways, and then all the laws of mind and matter that restrain it. Then the planets about it . . . and at last, out across immensity to the stars. And when he has conquered all the deeps of space, and all the mysteries of time, still he will be beginning.

PASSWORTHY: But . . . we're such little creatures. Our humanity is so fragile, so weak . . . little animals.

CABAL: Little animals [lowers his gaze to the ground] . . . and if we're no more than animals we must snatch each little scrap of happiness and live, then suffer and pass . . . mattering no more than all the other animals do or have done. It is this . . . [raises his gaze to the stars] or that. All the universe or nothingness! Which shall it be, Passworthy? Which shall it be?[97]

Here then, after a journey in literature across three centuries, we have a movie wearing the heart of the spaceflight and space expansionist narratives on its sleeve, filmed at a time when multiple private space societies around the world had organized, around ten years after Goddard had first flown a liquid fuel rocket, and in the same year as development had begun on the A-4 rocket—later to be known as the V-2. As the writer who spanned this era—and who spanned the spaceflight narratives' transition from literature to motion pictures—H. G. Wells deserves to be considered as one of the first great space philosophers. He understood that the process of stellar evolution would lead ultimately to the extinction of life on Earth, and he was the first to communicate this idea to the public. With an understanding of this future, he frames the choice before humanity as one between eventual extinction or space expansion—and he makes it clear that he believes that humanity should choose the latter. The period of the 1930s was thus not simply the prelude to the dawn of spaceflight in the 1950s and 1960s, as it is often portrayed in space histories. Rather, it represented the culmination of the development of the spaceflight narrative that had been evolving for hundreds of years and that was now prepared to engage the world with both a positive philosophy of space expansionism and the active development of rockets for traveling into space. In the closing lines of *Things to Come*, Wells had distilled all that had come before and dramatically asked the core question of spaceflight to the world.

With Wells, Serviss, and Goddard, we have interrelated individuals who each made their own contributions to the multigenerational development of a spaceflight culture. The spaceflight fiction of Wells and Serviss influenced Goddard, and Goddard's achievements and promotion of liquid fuel rocketry and space ambition influenced and impacted Wells and Serviss in turn. In many respects, the convergence of these three figures represents only one example of the codevelopment of spaceflight and spaceflight narratives, though a particularly significant and historic one. We see very much the same phenomenon today. Science fiction writers, space television actors, spacecraft developers, and space policy visionaries are all often in conversation with each other, often even know each other, with each codeveloping their contributions to spaceflight with the input and inspiration of one another. There have been innumerable such interactions of influence and encouragement in the development of interplanetary imagination and spaceflight ambition. This trio of visionaries, however—Wells, Serviss, and Goddard—reminds us both that the stories we tell can profoundly influence the future—by transferring knowledge, seeding new ideas, and encouraging passions—and that the question of what story we wish to write across the stars is one that we need to answer anew each day.

What story will it be, dear reader? What story will it be?

CONCLUSION

There's Nothing New / Under The Sun / But There Are New Suns.
—Octavia Butler, *Parable of the Trickster*, unfinished manuscript[1]

The spaceflight narrative today is one of the dominant technological narratives around the world. It has propelled humanity into space, and it continues to propel ever more ambitious interplanetary endeavors and interstellar ambitions as well. The spaceflight narrative has also proliferated extensively, with there being entire subgenres that seemingly produce more new fictional spaceflight journeys in a year than arose during the entire multicentury period examined in this book—and, of course, they do so across television, movies, and video games in addition to literature. There is extensive use of the spaceflight narrative outside of fiction as well. Today, spaceflight narratives are not just those described in books about fictional journeys to the Moon and interplanetary empires; they are also encoded in business plans, government reports, congressional hearings, and extemporaneous musings about the future of spaceflight on endless panels at space conferences. These narratives, which often purport to be factual descriptions of the world and the future, often also include significant amounts of fictionalization and projection. There are so many competing narratives about our future in space that the narrative landscape has become a potent and often bewildering mix, in which science fiction visions, national space policies, and space company strategies can become borderline indistinguishable. The ubiquity and power of the modern spaceflight narrative is one reason that it can be useful—in addition to entertaining—to examine the space narratives from earlier times in order to try to understand the recurring patterns, identify the genuinely novel ideas, and remember some forgotten ones.

We have seen how the spaceflight narrative began to develop shortly after the invention of the telescope. In a letter to Galileo Galilei, after the publication of *Sidereus Nuncius*, Johannes Kepler wrote about how the two of them could together establish the astronomy of the Moon and Jupiter in order to help prepare the way for the space settlers that he expected would someday make their way out to those worlds. In the late 1620s, after the public heralding of the telescopic discovery of a new world in the Moon by Galileo, the bishop of Hereford, Francis Godwin, wrote the first story to feature a human-built flying machine enabling a voyage to the Moon, *A Man in the Moone*—with its posthumous publication in 1638 marking the beginning of the modern spaceflight narrative as a literary genre. Shortly after its publication, in 1640, the bishop of Chester, John Wilkins, one of the founders of the Royal Society, wrote the first work of nonfiction about the potential for human voyages to the Moon, which he thought might well be inhabited, and suggested that a technology enabling spaceflight might open new opportunities for lunar trade and settlement. Cyrano de Bergerac had his protagonist attempt multiple different methods for reaching the Moon in his lunar voyage of 1657, although David Russen—writing across the English Channel in 1703 in Hythe, Kent—reviewed these methods in his own nonfiction treatise, *Iter Lunare*, and concluded that none of them would work, while nonetheless maintaining a belief that a voyage to the Moon might one day be possible, perhaps by means of a giant spring contraption built in the Himalayas. Two years later, Daniel Defoe would suggest that lunar extraterrestrials would use consumable fuel to power an engine of spaceflight in *The Consolidator*, and Murtagh McDermot would make the first suggestion of a gun launch system in a book first published in Dublin in 1727. The desire to conduct a space pilgrimage to the planets would appear in *The Life and Adventures of John Daniel* in 1751, shortly before Voltaire would weave his famous satirical tale of cosmic giants discussing philosophy while on interplanetary journeys in

Micromégas in 1752. By 1766, Marie-Anne de Roumier-Robert in France would suggest ideas for space travel that were evocative of later wormhole and teleportation concepts, before the invention and demonstration of the balloon in 1783 by the Montgolfier brothers would focus the attention of the next generations of spaceflight narrative writers on that first technology of flight.

One of these stories of balloon-borne spaceflight would be the first ever told in America, *An Account of Count D'Artois and His Friend's Passage to the Moon* by Daniel Moore, from 1785, which describes a spiritual journey through the heavens that is enabled by this new technology. In 1809, Washington Irving mentioned matter-of-factly, almost in passing, the idea that travel between the Moon and the Earth might be achieved as the result of further scientific advances—and that the invasion of one planet by another might be the result. George Fowler penned the second American visit to the Moon in 1814, which hinted at the spaceflight optimism to come with its moral elevation of the scientist and its utopian, cosmic perspective. In 1827, a former congressman turned professor resumed the serious consideration of spaceflight that had briefly begun in seventeenth-century England when he provided the first description of a spacecraft as consisting of a pressure vessel with an internal oxygen supply. Edgar Allan Poe provided spaceflight with a whole new sense of realism and verisimilitude in 1835 and also described for the first time the type of obsessive dedication that the dream of spaceflight could induce. In 1847, the People's Lyceum of New Orleans was witness to the return of a genuinely scientific consideration of spaceflight, for first time since John Wilkins, when the ingenious Dr. John Leonard Riddell gave the first-ever public lecture on the subject to a rapt audience. A scant four years later, in 1851, a quiet, young New York author provided the world's first description of a rocket-propelled spacecraft in a children's book.

In 1862, a Scottish minister and university principal in the United Province of Canada, William Leitch, wrote offhandedly in a religiously minded text that rockets would be a viable method for achieving spaceflight and thereby getting closer to God's glory in the heavens. Although Jules Verne's seminal *From the Earth to the Moon* was written in France, and published in 1865, it told a tale of how a distinctly American culture of optimism, private organization, and advanced industrial capabilities could combine to create a first voyage to the Moon. That story, along with its sequel *Around the Moon* in 1869, became one of the most important texts to inspire and transfer technical knowledge to the early pioneers of liquid fuel rocketry—most notably Konstantin Tsiolkovsy and Hermann Oberth. Also in 1869, and before Verne's book would be published in English, a widely respected Boston clergyman, who in his youth had seriously considered the development of a satellite system for determining longitude with his elder brother, published the world's first description of a space station and began in earnest the discussion of the space frontier as a place for experimenting with new social arrangements.

With the anonymous publication of *Man Abroad* in 1887, that potential had been wholly embraced, and the vision of space expansionism articulated, with an American republic extending throughout the solar system and with terraformed asteroids providing new opportunities for American individualism and radical political experimentation. The themes of the conquest of new planets, the subjugation of extraterrestrial cultures, and cosmic Christian salvation would merge in Florence Carpenter Dieudonné's incredible tale *Rondah* in 1887, while Cora Semmes Ives's earlier fable *The Princess of the Moon*, from 1869, introduced the theme of different groups of humans with different values entering into conflict on the Moon. In the 1880s, one of the founding figures of the Social Gospel movement, Washington Gladden, wrote the first story about a trip to the Moon that was accessible to middle-class ticket buyers, before the idea of spaceflight was given a new aura of credibility when New York high-society read John Jacob Astor IV's strange, visionary tale of interplanetary flight in 1894, foreshadowing the notion that the nation's multimillionaires might one day make spaceflight a reality. Finally, in 1898, *The Boston Post* would publish in serial two groundbreaking stories by H. G. Wells and Garrett P. Serviss that would inspire a young Robert Goddard to dedicate his life to the development of liquid fuel rocketry and spaceflight technology in general, and thereby bring to a close the multi-hundred-year era that saw the development of the concept of spaceflight proceed through literature alone.

The importance of spaceflight narratives to the development of space systems, and to the field of astronautics in general, continued throughout the twentieth century and continues up to the present day. When

Wernher von Braun decided to develop his concept for a crewed mission to Mars—the first such plan to be worked out in significant technical detail—he initially presented the concept not as a technical manuscript but in the form of a fictional narrative with an extensive technical appendix, the latter part of which would later be edited and published as *Das Marsprojekt* in 1952.[2] This work was not done as part of a government program, but rather on his own time and initiative while he was underemployed at the White Sands Missile Range in 1948, right after moving to America. Today, spaceflight narratives directly influence the composition of the workforce in our space industries, with numerous aerospace professionals citing twentieth-century science fiction as important determinants in their career decisions.[3] Once the narratives of spaceflight became historical as well as fictional—particularly after the Apollo program—these real-life narrative legends also played a similar role in generating interest in the spaceflight industry. With people currently living and working in orbit around the Earth, spaceflight narratives have continued to make important contributions by inspiring, transferring knowledge, and even influencing major personal investment decisions. Elon Musk—founder of SpaceX—has cited numerous spaceflight narratives as sources for his interest in space, including Douglas Adams's *The Hitchhiker's Guide to the Galaxy*, Isaac Asimov's *Foundation* series, Robert A. Heinlein's *The Moon Is a Harsh Mistress*, and Ian M. Banks's Culture series, and he even named his company's first launch vehicle—the *Falcon*—in a nod to the most famous spaceship from *Star Wars*. Jeff Bezos—founder of Blue Origin—was directly inspired by Gerard O'Neill's *The High Frontier* while in high school, and he has declared his love for Frank Herbert's science fiction epic *Dune*, named his personal holding company, Zephram, after the fictional inventor of the *Star Trek* warp-drive, and personally intervened to renew the science fiction television series *The Expanse*. Spaceflight narratives thus clearly continue to shape our spaceflight present, and they will certainly continue to shape our future in space as well.

What then are we to make of this motley collection of stories about spaceflight across hundreds of years, long before spaceflight—and even heavier-than-air flight—began? With such extensive historical and cultural breadth across these narratives—along with their incredible interplanetary and cosmic scope—there is no easy answer. I have written this book not because I think I have important insights into these stories, but because reading these spaceflight narratives, and thinking about them over the course of my career at NASA, provided me with new questions and thoughts regarding how and why we pursue our voyages into the vastness of the cosmos. There is no overarching thesis or critique motivating this work, other than simply recognizing that these stories and books were foundational to the conceptual development of spaceflight in the first place. My primary hope is that by collecting and summarizing them in one place, and by adding historical and biographical context, others can enjoy them, engage with them, derive their own conclusions, and perhaps embark on their own further explorations of the evolution and impact of spaceflight narratives over time. With those caveats aside, I shall try to offer my own perspective on what lessons we might try to learn from the spaceflight narratives that have been covered here.

The motivations described in the first few hundred years of the spaceflight narrative are diverse: Some revolve around conquest, some revolve around escape, some revolve around spiritual pilgrimage, some revolve around hope for a better world, and some revolve around existential species survival. Although there are no simple answers or conclusions, there is still much to be learned about how humanity became a spacefaring species in these early spaceflight narratives, through which we first articulated our aspirations for the stars and how we might reach them. The very diversity of motivations in these early spaceflight narratives is, in many ways, perhaps the most important conclusion. There is no one reason that motivates us to travel into space, and there is no one right answer for the question, "Why space?" For some, the planets represent new, near-mythological lands on which to establish new empires and sources of vast wealth. For others, the planets represent opportunities for dramatic and dashing adventures, while others still see them as opportunities to advance human knowledge through scientific investigation. For yet others, the planets represent guide stones on a path to knowing the divine, and visiting them is more akin to a pilgrimage to Jerusalem or Bodhgaya or the Oracle at Delphi than it is to the voyage of Columbus across the Atlantic. For others still, spaceflight represents the ultimate option for the indefinite continuation of intelligent life through interplanetary and interstellar migration. For many, the interest in a future that includes further extensions of spaceflight is motivated by a varied and heterogenous mix of some or all of these motivations, or other ones altogether. In the longer historical context, from a narrative perspective, none of these radically different perspectives on the value and importance of spaceflight has clear priority or dominance. In order to accommodate this genuine diversity of motivation and ambition, we should perhaps adopt and accept a wider diversity of policy objectives and mission destinations within our global space community, and focus more clearly on the enabling capabilities that allow a wide array of these narratives to be explored and experimented with simultaneously.

There is wide diversity in the motivations observed across the centuries examined here, but not all motivations are observed equally across time, and some motivations are observed earlier than others. For example, the very first motivations expressed are, interestingly, those of settlement and trade, as seen in the writings of both Kepler and Wilkins. Interests in and expectations of space settlement thus date back to within only a few years of the founding of Jamestown in 1607. Individualistic adventure and spiritual interests were also among the most prominent early motivations depicted for spaceflight across the seventeenth and eighteenth centuries, as they were for Domingo Gonsales, John Daniel, Count D'Artois, and Randalthus. The earliest stories of spaceflight are predominantly stories of individual, personal adventures undertaken out of idiosyncratic and intrinsic interests. It is in the nineteenth century that broader social motives begin to be introduced, with Verne's Baltimore Gun Club conducting its mission to the Moon as a public performance and with John Jacob Astor IV similarly imagining that the launch of his protagonists to Jupiter would be a grand public spectacle. Nationalism and internationalism also become motivating sources of support in the nineteenth century, with *From the Earth to the Moon* and *Man Abroad* depicting spaceflight as a particular enthusiasm and achievement of the United States of America and *Edison's Conquest of Mars* describing the first international spaceflight coalition forming around the need to eliminate the threat that Martians pose to Earth. Although depicting different political systems had been part of imagined lunar voyages from the start, it was largely in the context of imagined extraterrestrial societies on the Moon. It was not until the nineteenth century that the idea of humans moving into space for the purpose of political experimentation becomes a significant theme. In tracing the evolution of these spaceflight themes over time, it is striking that spaceflight began in stories with overridingly personal and individual contexts, but that many of the most significant social and political elements of spaceflight and space expansionism nonetheless were also already described in the spaceflight narratives of the nineteenth century.

It is also something of a surprise, when reading through these stories, to note the relative rarity of themes of conflict and war. Given the ubiquity of space battles and proverbial "star wars" in more modern spaceflight narratives, one might expect these themes to have been present from the start. With both Godwin and Wilkins writing their texts while the devasting Thirty Years' War raged on the European continent, and Daniel Defoe writing his spaceflight narrative at the time of the War of the Spanish Succession, it would not have been surprising if these spaceflight narratives had included martial themes, but it wasn't until the nineteenth century that they came to the fore. Washington Irving would not describe the potential for an invasion of Earth by philosophers from the Moon until the 1800s, and the idea of different groups of humans entering into conflict on another celestial world doesn't appear until Cora Semmes Ives's story in 1869. True interplanetary war, with large-scale death and destruction as major thematic elements, is not a significant focus for a spaceflight narrative until *The War of the Worlds* and *Edison's Conquest of Mars* in the 1890s. For some two hundred years, then, we managed to tell stories of spaceflight without depicting it as an opportunity for new expressions of warfare. We might want to consider more thoroughly reviving this older tradition of spaceflight storytelling.

Another aspect of spaceflight storytelling with a relatively late appearance is the notion that we may need to move off of Earth in order to mitigate existential risks to the survival of the species. H. G. Wells, with his understanding of both biological evolution and stellar evolution, introduced this idea first in *The Time Machine*, then again in *The War of the Worlds*, and then most unequivocally and poetically in *Things to Come*. Again, it is interesting both that this particular argument for spaceflight, which is a core motivator for many space expansionists, has been around for well over a hundred years, and also that the broader spaceflight narrative had existed for around 250 years before it emerged. This existential argument for space expansionism has today become one of the most commonly cited rationales for the necessity and importance of spaceflight, and yet the idea of spaceflight itself was able to develop and evolve very successfully without it for hundreds of years. The vast majority of the spaceflight narratives covered here were depicted as having more personal motivations—which we might call the *experiential arguments* for space expansionism—based on the specific experiences that the individuals undertaking the journey had, or were expected to have. Although the existential argument for spaceflight is a strong one in the final calculation—as it will indeed eventually be either "all the universe or nothingness" for humanity—that final calculation remains a very, very, very long way off. As this journey through the earliest spaceflight narratives reminds us, a lot can change in a few hundred years, let alone in the roughly one billion years that we have until Earth is eventually made uninhabitable by the expanding Sun. In addition to recognizing the economic and strategic significance of space technologies, we might also want to give higher priority to the experiential motivations for the spaceflight enterprise and to shape our space policies accordingly, at least for the next few thousand years.

That we have been thinking about building spacecraft for almost four hundred years is also an important reminder that, as special as our particular moment in spaceflight development may seem, we are in fact only the latest generations in a long line of spaceflight dreamers and doers. If we take the publication of Godwin's *The Man in the Moone* in 1638 as the starting point, and we assume that a generation is roughly thirty years in length, then we are nearing the beginning of the fourteenth generation of creators of—and, to varying degrees, believers in—the spaceflight narrative. We are inheritors of a vast amount of thought and labor that has helped pave the way for the incredible abundance of spaceflight capabilities now at our collective disposal. In addition to the multi-century-long history of the concept of spaceflight itself, there are also multi-century-long histories of space expansionism and of specific technical ideas—such as those for spacecraft and spacesuit life-support systems, orbital space colonies, in situ resource utilization, and terraforming, all of which first emerged in the nineteenth century. Our modern spaceflight institutions—national space agencies, space companies, private space societies—are all less than one hundred years old, and so it often feels like spaceflight communities and stories are recent phenomenon. In fact, the first spaceflight communities were intergenerational, literary ones, and they have existed for hundreds of years. We should not forget this longer history, nor the men and women who helped us arrive at the present moment through their works across it, as their efforts were in their own way as vital to our present moment as were the efforts of the first spaceflight engineers and program managers of the twentieth century.

In reading these spaceflight narratives, I am also struck by how we have yet to meet, let alone surpass, many of the visions of the nineteenth century. We still do not have people who permanently live out their lives in low earth orbit, over 150 years after "The Brick Moon." We still do not have asteroid settlements, around 150 years after *Man Abroad*. We still have not sent humans to Jupiter or Saturn, even though new billionaires have taken up John Jacob Astor IV's mantle over a hundred years after *A Journey in Other Worlds*. In fact, with the exception of orbital spaceflight and human voyages to the Moon, the expansive space futures described by writers in the nineteenth century remain largely unrealized, even though there have been major advances in technology and many continue to labor at bringing versions of these visions into reality. Many continue to work toward these visions through academic research, ambitious space companies, and aspirational space policies and programs—and yet, in some cases, it is legitimately unclear if we are closer in time to the realization of these visions or to the time when they were first written about in the nineteenth century. That we are still working on trying to bring into reality ideas from that long ago should perhaps give us pause to consider how soon some of these ideas may, in fact, actually be realized—if indeed they ever will be at all. We still await the space settlers that Kepler believed would be forthcoming. We are nearing four hundred years of anticipation. Might we still wait a few hundred more?

As my old dissertation advisor, Avner Offer, would often say: The only problem with stories is that they are too convincing. Narratives can readily become more persuasive than objective analysis in assessing potential space futures. I have seen this repeatedly over the course of my career at NASA. This understandable human bias toward persuasive narratives can pose real risks to program management, not to mention managing public expectations. Although this book is a celebration of wild and inspiring spaceflight narratives, it is also a reminder that compelling spaceflight stories—whether those found in old books or those found in the dazzling presentations of enthusiastic space expansionists—provide no guarantee that they can be realized. We have indeed gone to the Moon, much as Verne described we would, but we did not actually use a cannon to do so, to say nothing of Poe's carefully detailed balloon or Godwin's geese. Our stories of spaceflight may hold important truths, and inspire real projects of broad economic and social benefit, but they may also be misleading in significant ways, and not a few spaceflight stories are simply pure fantasies. This can be as true for our purportedly nonfiction spaceflight narratives as it can be for our fictional ones. Although there are no limitations to the spaceflight stories we can tell, there are physical constraints in the universe that remain immune to hype, and there are limitations to the machines we can build at any given time. We should remember this when we find ourselves rising in excitement for inspiring spaceflight narratives that have, in some cases, remained unrealized and remarkably unchanged for decades—and, in some cases, for hundreds of years.

What is there new for us in these stories, then, with physical spaceflight now in its seventh decade, and many of these themes familiar to us from popular culture and the pitch decks of space entrepreneurs? Perhaps it is from the rarer, forgotten stories that we have something to learn. Personally, one of the most interesting findings for me in writing this book was that the idea of space pilgrimage has a history that dates back hundreds of years. In fact, space pilgrimage is the motivation of not only the first but two of the first three of the very earliest examples of American space narratives that have thus far been found. It is also implied by John Wilkins to be a motivation for voyaging to the Moon in the first nonfiction book ever written about spaceflight. We can even think of the final lunar journey of the Hindu scientist-priest in George Tucker's *A Voyage to the Moon* as being another type of space pilgrimage or, in his case, a *gaganyatra*. Although we don't often think of pilgrimage as a spaceflight motive, it is perhaps not as uncommon as one might think. Arguably the most memorable single moment of the Apollo 8 mission was the reading from Genesis by the astronauts while around the Moon, and Buzz

Aldrin, as part of the first human mission to the lunar surface, also took time to take the Eucharist. On a more personal level, I have undertaken a number of terrestrial space pilgrimages myself—and explicitly thought of them as such—to astronomical observatories, important launches, and other sites of significance in the history and development of our study of, and movement into, the cosmos. It seems inevitable to me that, if our capabilities for spaceflight become sufficiently affordable, accessible, and advanced, at least some of us will choose to make spaceflight journeys as pilgrimages. Indeed, as mentioned earlier, it is arguable that we already have. If we seek a spiritual meaning in the heavens—and what human culture has not looked to the stars and planets in search of meaning?—then we might consider more explicitly thinking about the types of pilgrimages we might make in our solar system.

A modern reader of these early spaceflight narratives also cannot help but be struck by how one of the most common and pivotal elements of many of the stories is some sort of encounter with extraterrestrial life. In some respects, this is not entirely unexpected as aliens and imagined alien encounters continue to be popular in space narratives today. But what is striking is that during the seventeenth to nineteenth centuries, these alien encounters were all expected to be much closer to home and much more readily encountered than has turned out to be the case. With the Moon expected to be both habitable and inhabited by leading scientists in the seventeenth century, somewhat paradoxically these scientists held a higher expected likelihood for interplanetary emigration and settlements within our solar system than most scientists do today. We are, in some respects, still grappling with the fact that our solar system has turned out to be far less inhabited, and even far less habitable, than had been expected during the first centuries of the development of the spaceflight narrative. Our robotic space probes have shown us incredible images of strikingly beautiful worlds, but none of them have yet shown us extant life, let alone alien civilizations, as many of these earlier spaceflight narratives had hoped would be readily discoverable. Indeed, in some respects the spaceflight narrative was never the same after the first flyby of Mars by Mariner 4 in 1965, which revealed a seemingly dead, crater-marked planet after centuries of hopes and stories focused on nearby extraterrestrial civilizations. This has created a challenge. A popular belief in the spaceflight narrative and space expansionism remains strong, but one of the primary, originating motivations—the appeal of encountering diverse extraterrestrial life and alien civilizations only a relatively short interplanetary flight away—has effectively evaporated.

Furthermore, the more we have progressively learned about the planets and about human biology in space, the less it has seemed likely that we should expect to be able to readily expand into space in large-scale numbers any time soon. Although many have laid out the reasons for this over the years, Kelly and Zach Weinersmith's book *A City on Mars* does an excellent job of explaining why this is the case.[4] This may, paradoxically, actually highlight the continued importance of spaceflight narratives overall. Compelling spaceflight narratives will continue to be needed if we are to maintain the faith that humanity can yet, some distant day, have an expansive future in the heavens. These stories can continue to help identify solutions to some of the most difficult challenges and to inspire potential inventors and innovators to solve them, just as spaceflight stories have in the past. Because we are nearing four hundred years of anticipation for expansionary space settlements, it is not inconceivable that we might have to settle in and prepare patiently as we work toward them for a few hundred more. We might therefore still look forward to a future in the stars, but we might not live to see true space settlements—with families and natural increase—in our solar system for many decades, centuries, or even perhaps ever. If that should be the case, is there any reason to still value and believe in the spaceflight narrative or space expansionism?

Although not regularly thought of as one, Octavia Butler is, for me at least, one of the most important philosophers of space expansionism in the twenty century—in particular, through the spaceflight philosophy that is at the heart of her books *Parable of the Sower* and *Parable of the Talents*, as well her unfinished third book, *Parable of the Trickster*, and the three other books that Butler had planned to finish the series: *Parable of the Teacher*, *Parable of Chaos*, and *Parable of Clay*.[5] Butler's *Parable* series starts in a terrifying and devastated American future, one of regular violence and ecological collapse. The originating protagonist of the story, Lauren Olamina, navigates through this apocalyptic future with her idiosyncratic religious belief system, Earthseed, which she believes she must spread throughout humanity so that humanity can expand to the stars. The vision—"just two lines," as she says—comes to her one morning, July 20, 2025:

The Destiny of Earthseed
Is to take root among the stars.

This is what I was reaching for a few days ago when the story of the new planets being discovered caught my attention. It's true of course. It's obvious.[6]

Olamina then proceeds with a fairly realistic perspective on how far away this vision remains, which it is hard not to find resonance with today:

Right now, it's also impossible. The world is in horrible shape. Even rich countries aren't doing as well as history says rich countries used to do. President Donner isn't the only one breaking up and selling off science and space projects. No one is expanding the kind of exploration that doesn't earn an immediate profit, or at least promise big future profits . . . I don't know how it will happen or when it will happen. There's so much to do before it can even begin. I guess that's to be expected. There's always a lot to do before you get to go to heaven.[7]

For Olamina, space expansionism and the interstellar spaceflight narrative is a potential source of hopeful faith, even—perhaps especially—in the most difficult and trying of times.

In *Parable of the Talents*, Butler provides one of the most poetic descriptions of the space expansionist creed:

The Destiny of Earthseed
Is to take root among the stars.
It is to live and to thrive
On new earths.
It is to become new beings
And to consider new questions.
It is to leap into the heavens
Again and again.
It is to explore the vastness
Of heaven.
It is to explore the vastness
Of ourselves.[8]

Butler's Earthseed is similar to H. G. Wells's vision of a seed from Earth spreading across the cosmos, and it is in many ways a distillation of space expansionism, boiled down to its core. In the first two *Parable* books, there are few narrative references to spaceflight other than this belief and the mention of ongoing missions to Mars, and yet at the end of *Parable of the Talents* the ultimate success of Earthseed is made clear with the launch of the first interstellar transport ship carrying settlers to a new world around another star. *Parable of the Sower* and *Parable of the Talents* take place in a cataclysmic and brutal version of America and the world, and the belief system of Earthseed provides a source of hope for Olamina when all other sources of hope are lost. This, for better or for worse, may be what we are left with. There may be nothing new under the sun, but there are indeed new suns—new suns that it remains just barely possible to seriously imagine traveling to, and this alone can provide new sources of hope, no matter how faint or distant.

If the challenges of expansion within our own solar system are daunting, however, then the challenges of interstellar travel and expansion are vastly more forbidding still. Adding thousands of years—or tens, or hundreds of thousands of years—to our prospective timeline may allow for as yet unimaginable advances in science and technology to enable us to overcome these nearly incomprehensible challenges. Then again, it is far from inconceivable that those challenges might prove to be simply insurmountable. If that is the case, if we are, even in the final account, confined to the planet or solar system of our origin, what then is the value or purpose of the spaceflight narrative or a belief in space expansionism? In addition to being a story about humans building machines that can allow for adventures, resource development, and scientific discoveries beyond Earth, the spaceflight narrative is also a story of possibility, of potentiality, of seemingly impossible dreams that are in fact realized, and of how we might change as a result of their realization. It is, in short, a story of who we might yet become. By believing in and working toward the systems that enable us to travel into the heavens, not only have we become a people who can and do, but we have also built a world with an orbital infrastructure that provides a greater ability than ever before to communicate with each other, navigate our world, and understand our planet and our universe. As we have worked toward expansive spaceflight visions, we have built the capabilities and technologies for a more expansive life on Earth. The stories in this book, across four hundred years, were part of that process, but the stories that result from our spaceflight achievements in our solar system over the next four hundred years will likely have an even greater impact.

This is, in part, because the cosmos is the grandest stage imaginable for the expression of our humanity. As we proceed with our flights out to the planets and beyond, all the world gets to observe and consider who we are becoming in the process, and the values that are expressed and embodied in that process will be of historic significance. We have seen this already with the Apollo program, which famously "came in peace" for all humankind, and with the International Space Station, which united multiple countries around the planet and showed that even former adversaries could work together to build, and then work and live in, humanity's first international home in space. Each major new spaceflight narrative that we pursue and make real—whether a permanent Moon base, a first human mission to Mars, or a planetary defense system to protect ourselves against asteroid impacts—will matter not just because of what is achieved or gained through it but also because of the stories that emerge from it and what they reveal about who we are, what we value, and the future we are trying to bring into reality.

What then can we make of space expansionism and its early historical roots? If space expansionism is indeed a belief system, then in some respects the real debate is over the values and core beliefs that we should hold and uphold as we expand into space—and the history of the spaceflight narrative shows as much diversity in this regard as we see today. Because a potential future in space for humanity is seen as so expansive, many believe that the values that are established now matter a great deal, as we are in theory at the start of that expansion and many wish to see their values propagated. This is behind much of the argument within US space policy circles for why the US needs to return to the Moon—and stay there—before China does so, due to concerns that otherwise non-American values and interests might become more prominent throughout the rest of our future in space than American ones.[9] This is also behind much of the critique of space expansionism that has emerged, with some arguing both that we should temper the expansionist impulse and that ecological values should have greater prominence within global space policies and preferences.[10] The nineteenth-century space narratives featured in this book implied debates of arguably even greater polarity. The history of the spaceflight narrative includes visions of dystopian lunar "utopias" and brutal interplanetary conquests. But it also includes visions of planetary and interplanetary harmony as well as international and interspecies cooperation. Florence Carpenter Dieudonné's *Rondah* celebrated the righteous power of a single man to become a divinely ordained cosmic monarch, while H. G. Wells argued that the future of humanity in the cosmos necessitated a true world state managed by scientists. In the debate over the values that should guide our expansion into space, we would do well to remember both that this debate is far from new and that, correspondingly, it is unlikely to end soon either. There are horrors and cruelties in some of our spacefaring narratives. There are also beauty, generosity, and love in some of them as well. Our visions of a future in space are, in short, thoroughly human. Space expansionism—just like spaceflight technology itself—is neither inherently good nor inherently bad. At its core, it is simply the belief in a story of the future that includes humans experiencing some part of life off Earth. We should both recognize the power of that story and continue our inherently human debate around the values that we should enact and uphold as we seek to expand into the cosmos.

The spaceflight narratives described in this book, however, should also serve to remind us not only that there are many different spacefaring values and different space communities, but also that in some respects there is a core spaceflight narrative that is sometimes overcomplicated by the grand visions often embraced by the global space community. Within the US space community in particular, the evolution of the spaceflight narrative has arguably led to there being distinct "space tribes," each with different values and narratives at their core. One popular framework depicts three competing and contrasting visions for the future of humanity in space, each one associated with the worldview of a major space expansionist thinker from the twentieth century. In this framework, there are Saganists, Von Braunians, and O'Neillians.[11] Saganists, named after Carl Sagan, purportedly work toward a vision that prioritizes scientific exploration and environmental ethics in space activities. Von Braunians, named after Wernher von Braun, purportedly work toward a vision that prioritizes national power and prestige in space activities. O'Neillians, named after Gerard O'Neill, purportedly work toward a vision that prioritizes space settlement and space commerce in space activities. Although the framework is reductive—and is predominantly used by self-described O'Neillians—it is a sometimes useful shorthand for policy and narrative preferences, which can enable discussions on how a particular policy option might be seen through the lens of different interest groups. It is important to note, however, that all three are space expansionist narratives at their core; they differ not in any fundamental belief in the value of any movement out into the cosmos, but rather in what the priorities and motivations for such a movement should be. The stories in this book might remind us, however, of the originating motivation that we see on display in many of them: spaceflight as a simple expression of personal, idiosyncratic human interests.

Although there were many different elements of space expansionist thought in the first stories of spaceflight, it was the core spaceflight narrative—that humans can build machines that travel into space, and thereby have unique experiences and new adventures—that preceded and ultimately gave rise to the variety of space expansionist futures now being propagated and debated. We might do well to remember and respect the simplicity of this ur-narrative of spaceflight, as it was this simple narrative thread that dominated the conceptual development of spaceflight for its first couple of centuries. One might even argue for the value of identifying an original "space tribe," one we might call *Vernians*.[12] Vernians, named after Jules Verne, worked—and in some cases, still work—toward a vision that prioritizes idiosyncratic personal experiences and adventures in space activities. A Vernian space expansionist might, for example, eschew a likely quixotic attempt to develop a near-term colony on Mars in favor of a significantly safer and more realizable human mission to flyby or orbit the planet Venus, simply on the grounds that doing so would be a beautiful and personally meaningful adventure. A Vernian space expansionist might believe that no large, self-sustaining human settlements are likely to develop in our solar system for the foreseeable future and that interstellar spaceflight is likely centuries away or more, and yet also believe that human spaceflight is nonetheless worth pursuing on account of the inspirational and cultural merits of the enterprise. A Vernian space expansionist might see the Artemis II mission as one of the most profoundly important space exploration missions in history based solely on the feelings of joy, wonder, and hope that it rekindled in so many people across the planet. There is something to be said for valuing the experiential arguments for space expansionism at least as much the existential ones, and for celebrating the fact that humans are able to leave the surly bonds of Earth at all, have unique personal experiences in the cosmos, and return to Earth to tell us about them so that we can reflect and learn from their experiences. Perhaps, in appreciating and centering a simpler spaceflight narrative, we might find our space present as exciting as any imagined, unrealized space future.

The spaceflight narratives that we prioritize are important not only for the values and philosophical lenses that they offer, but because these narratives also represent a type of cumulative labor that, when combined and successively built upon, shape the future that we ultimately experience. The spaceflight narrative overall has helped to lay the foundation for the real spaceflight systems that have emerged and on which we continue to improve. Spaceflight narratives have thus served—and still serve—as vehicles for knowledge transfer and as sources of motivation for the personal labor-allocation decisions that provided critical, early resources in the development of spaceflight technology. In the basic manner described in Robert Schiller's *Narrative Economics: How Stories Go Viral and Drive Major Economic Events*, spaceflight was a narrative that started going viral in the seventeenth century, and it has only grown in its import and influence ever since.[13] It has certainly driven major economic events, including the allocation of massive amounts of US federal spending to pay for the Apollo program and the rapid rise of Elon Musk to the status of one of the wealthiest individuals in modern history—with both of those phenomena resulting, at least in part, from the power of the stories told surrounding them. Which types of spaceflight narratives we choose to elaborate and develop, therefore, might have a real impact on the type of future we live in.

In economic terms, I would argue that the intrinsic motivation that results from belief in the spaceflight narrative creates a type of kinked labor-supply curve—with a sufficiently strong belief resulting in an amount of labor being supplied, and therefore a certain amount of production realized, independent of market prices for this labor and regardless of whether or not there is demand. This is the core power of the spaceflight narrative—and narratives in general: the ability to motivate labor irrespective of external reward. Spaceflight narratives can even be considered a type of input for spaceflight production. Konstantin Tsiolkovsky, the preeminent Russian theorist of early spaceflight and one of the inventors of liquid fuel rocketry, summed up the role of spaceflight narratives in the process of spaceflight production in the quote presented at the beginning of this book: "Science-fiction stories on interplanetary trips carry new ideas to the masses. All who work on these lines work well, arousing interest, promoting the working of the brain, and training up people who are attracted by the idea of grand projects, and who will work in this field in the future."[14] Tsiolkovsky spoke from personal experience as both a receiver and transmitter of the spaceflight vision. It was Verne's *From the Earth to the Moon* that instilled knowledge and motivation in both him and Hermann Oberth, the German inventor of liquid fuel rocketry. Tsiolkovsky's work influenced, in turn, the Soviet rocket genius Sergei Korolev, while Oberth's *Die Rakete zu dem Planetenräumen* inspired von Braun to take up the serious study of mathematics at the age of thirteen.[15] A line of technical influence and inspiration runs from the two leading figures of the Cold War space race programs—Korolev and von Braun—through Tsiolkovsky and Oberth, through Verne and Poe, all the way back to Godwin and his gansa engine spacecraft of 1638.

The development of liquid fuel rocketry thus sprung from the intrinsic interest of individuals who were driven by a particular belief in the future and who were equipped with enormous dedication and persistence—traits that had also been identified as important enablers in several of the first narratives describing fictional spaceflight heroes. The early life decisions of Goddard, Tsiolkovsky, and Oberth to dedicate themselves to the problem of spaceflight resulted in them collectively devoting some 190 years of highly skilled labor.[16] Moreover, they were hardly alone. Influenced by science fiction—as well as by the work of earlier liquid fuel rocketry pioneers—Wernher von Braun, Arthur C. Clarke, Eugene Sanger, and Krafft Ehricke would make similar decisions in their teenage years to dedicate their lives to the pursuit of spaceflight: a further two hundred years of dedicated skilled labor.[17] It is not too much of an exaggeration to say that the spaceflight narratives of the past not only motivated but also built the first foundations of our own spaceflight present. This cycle of new ideas, inspiration, and motivated labor continues. Spaceflight itself is narrative economics in action.

The spaceflight narratives of the preceding centuries have provided both motivation and knowledge for the generations that followed. Just as importantly, starting in the early twentieth century—with the formation of rocket societies in America, Britain, Germany, and the Soviet Union—individuals who believed in the spaceflight narrative began to find each other and form groups and communities—first in private societies, and later through companies and government agencies. As much as the myth of the lone inventor is celebrated in many of the earliest spaceflight narratives, by the nineteenth century there was an increased understanding of the economic and technical complexity of spaceflight, and so the role of teams and groups became much more prominent in spaceflight storytelling. Indeed, in the case of Verne, he even imagined a private society—much like those that would later be formed to pursue spaceflight—as the locus of spaceflight realization. Were we only our limited individual selves, we would never be able to voyage into the cosmos. But in groups and organizations, working to high ideals, it becomes possible to build the machines and the world that make real even interplanetary voyages. Spaceflight narratives did not only inspire individuals: They inspired societies, eventually whole nations, and arguably the world. Along with other faith-based narratives, one might argue that spaceflight narratives are among the most powerful stories ever told. Indeed, spaceflight narratives can be so powerful that one might become the richest man in the world in part by harnessing, selling, and delivering on a good one.

With the emergence of a global spaceflight culture, some of us have even placed the spaceflight narrative, and the philosophy of space expansionism, at the mythic center of our modern lives. In a world in which science and technology have shifted the fulcrum of spirituality for many, the existential implications of spaceflight can seem to provide answers for some to life's enduring questions around identity, vocation, purpose, and ultimate ends. Although H. G. Wells would outline a collective existential argument for spaceflight from the perspective of humanity as a whole in the 1890s, Edgar Allan Poe had already made a personal, psychologically existential argument for spaceflight over half a century earlier in the 1830s. Edgar Allan Poe depicted his tortured Hans Pfaall as desiring to travel into space only after despairing, and giving up on his life on Earth, with the possibility of undertaking a voyage to the Moon serving as his personal answer for why he should continue to exist at all. It was an internal story of spaceflight, a personal narrative that came to him after reading popular astronomy books, that gave his life new energy and momentum. Poe's story is arguably the earliest to have identified the transformational role that the spaceflight narrative can play within an individual psyche. Edward Everett Hale also considered the role that spaceflight might play in answering life's existential questions, albeit more in terms of its potential role in shared social identity and communal organization. Through their new lives on the brick moon, the multifamily Sandemanian community finds new collective purpose and more harmonious interpersonal relations in orbit than they had enjoyed on Earth. Hale describes a community venturing into space and finding that they can create a new and better life story for themselves as a result. Spaceflight is thus presented, even in the nineteenth century, as a means of existential salvation for the individual, for a community, and for humanity as a whole. We thus have over 150 years of experience projecting our desire for new beginnings and better versions of ourselves out onto imagined new lives in space.

This leaves us with a critical question that is raised by this exploration of the original spaceflight narratives: What new spaceflight narratives should we create? For many decades, it seemed that the global space community believed in, and was working toward, a *Star Trek*–style future in space, one in which individuals from many nations would cooperate peacefully in the pursuit of science and exploration in the vastness of the cosmos. Increasingly, however, it seems that the global space community is shifting toward believing in, and working toward, a *Dune*–style future in space, one in which all that matters is the competition between the Great Houses for power, control, profit, and resources across the galaxy. Although I confess that I enjoy engaging with and thinking about both of these types of spaceflight narratives, I nonetheless know I would personally much rather live in a *Star Trek* future than a *Dune* one.

At the moment this is being written, therefore, there seems to be a need to bring forward new narratives as well as to reclaim old ones: narratives of building peace and global cooperation in space, even among rivals; of humanity not only avoiding a space war but of using space to avoid and reduce conflict on Earth; of using space systems to improve the management of our planet and our lives on it; of space economies with mutual aid and mutual plenty; of new space adventures, of idiosyncratic experiences, of human beauty; of learning to navigate our wild, technological civilization long enough to be able to last the hundreds—perhaps thousands—of years that may be needed until we have a chance of truly traveling to "new Earths" around other stars; of wandering space pilgrims flying by planets for the simple joy and spiritual exaltation to be had in doing so.

In addition to fostering and integrating new spaceflight narratives, we would also benefit from further research into the role that storytelling about spaceflight has played in early space expansionism, popular culture, the earliest space businesses, and the origins of space policy and governmental space program development. This book has attempted to cover a significant amount of geographic and temporal ground, and yet in truth it is barely a beginning. I have no doubt that others interested in the literary history of spaceflight and the intellectual history of space expansionism, and who dig deeper into the contexts of different writers and thinkers, will find many important connections and interpretations that I have missed. The nineteenth century alone merits deeper research into at least the English, French, Russian, and German spaceflight narratives, along with those from other parts of the world that may have also experienced intellectual spaceflight flourishes during that remarkable century. The intellectual history of space expansionism in the first half of the twentieth century has not yet been systematically investigated, even though it was during that period that the ideas of the spaceflight narrative led to the formation and growth of the first rocket societies and began to be propagated ever more extensively through pulp magazines, comics, radio, film, and television. Although there have been excellent histories of some of the early rocket societies written, I have spent long enough in the archives of the British Interplanetary Society to know that at least one good doctorate awaits an individual willing to thoroughly search its earliest records and trace the development of early twentieth-century spaceflight thought there, and I suspect the same is true for other early space societies around the world. Although there is no shortage of literary and cultural analysis on twentieth-century science fiction, the analysis of science fiction in terms of its relationships to, and feedback systems with, the development of space expansionist thought is still likely fertile soil to till. As spaceflight itself began to be achieved and regularized, spaceflight cultures proliferated around the world in the twentieth century, and we would benefit from literary and intellectual histories of spaceflight and space expansionism in Japan, India, China, the UAE, Saudi Arabia, Australia, Canada, and many other countries. Finally, it is beginning to become feasible to productively use natural language processing methods to analytically assess economic characteristics associated with narratives, and we await the first successful application of this powerful new method to narratives related to spaceflight.[18] Stories and ideas are at the root of our voyages into space. It is time that we became serious about studying their history and their impact on our societies.

The history of the origins of the spaceflight narrative should remind us not only that there are consequences to the types of stories that we create and promote but also that there is far more diversity to be had in our spaceflight storytelling than we have commonly realized. Spaceflight is the result of a story about a future for humanity in space in which enough people have believed, and for which enough have used their labors and resources, to make spaceflight a reality. The belief in that story—in that hypothetical future—is at the core of our flights out into the cosmos. Spaceflight is, in this sense, a faith-based enterprise. It is pursued, and often led, by people who have faith in the story that there will be a future for humanity in space—or at least, who believe that that story is a more appealing and hopeful one than a story where there is no future for humanity in the cosmos at all. For those that do not believe in that story, hopefully this book has provided some insight into where that belief comes from. For those of us who do believe in that story, let us understand that many will—with good reasons—not believe it at all, and others will believe it for different reasons than we do. Let us nonetheless seek to find common ground and work together to build new stories—both fictional ones and ones written through our actions and adventures beyond Earth—that inspire others, decades and centuries into the future, to find their own new reasons for believing that humanity may yet have a future among the stars.

Perhaps the most fundamental lesson, however, is simply that the process of creating and telling stories can motivate and inspire the ultimate realization of those narratives themselves. These stories serve as a reminder of the power of narrative—and indeed, of books—to shape the world centuries into the future. Spaceflight narratives are everywhere around us in the space industry and in the world—but how often do we stop and ask ourselves, "What future are we bringing into existence through our stories?" In addition to recognizing the importance of the stories and people that first developed the concept of spaceflight, we need to remember the importance of the stories that we tell each other about the future more generally. These stories don't just transmit information or ideas; they shape beliefs and nurture passions, beliefs and passions that can lead us to dedicate our lives to the realization of important projects. Stories can and do influence social and technological forces centuries into the future. Today, more than ever, we need to realize this and remember it when we both create and consume our narratives. We need to work hard to develop and amplify narratives that don't just show us the possible dystopian paths we may take in space—for not only do such paths seem concerningly plausible, but also such dystopian stories can plant seeds for possible dystopian futures. We may instead want to develop a bias toward telling stories that plant the seeds for, if not necessarily utopias, then at least new hopeful projects of economic, societal, spiritual, institutional, and technological transformations.

If we think that this idea—that the narratives we tell each other can transform the future—is fanciful, or impossible, then we need to remember the story of the spaceflight narrative: a story from the seventeenth century that propagated culturally for over three hundred years until it could finally be realized. Spaceflight is, at the end of the day, a story so motivating that, after telling it to ourselves for hundreds of years, we finally made it happen. This may, in fact, be its most beautiful and enduring characteristic in the life of humanity.

ACKNOWLEDGMENTS

The story of this book is, in a way, its own strange little spaceflight narrative. It began over a decade ago as a chapter in the first version of my doctoral dissertation on the economic history of space exploration. My doctoral examiners, however, were less convinced than I was that early spaceflight narratives should truly be considered part of that history. The chapter was thus removed from the work that would become *The Long Space Age*. I would later circulate the chapter with a few friends and colleagues interested in the topic, but it would largely lie undisturbed for several years across multiple hard drives as I proceeded with my work at NASA as the agency's first chief economist. However, I never could shake the thought that the stories we create and consume about spaceflight are of fundamental importance in the overall production process that leads us to build spacecraft and travel into the heavens. I gave a TED Talk in 2019 about the basic elements of this process, but I still felt the need to try to explore and explain its history more fully. The result of that unshakeable interest, now many years later, is this book.

As I came to finish this work, I began to recognize that it is as much a love letter to books as it is anything else. My own interest in the possibility of a human future in space began with stories, and with books in particular: nonfiction books about the stars and the planets, but also books about fictional space adventures. The first spaceflight narratives I can remember reading and loving are not any included here, nor ones that are regularly considered among the classics of science fiction, but rather the earliest books of the *Star Wars* expanded universe; Timothy Zahn's *Thrawn* trilogy; and the short story collections *Tales from Jabba's Palace* and *Tales of the Bounty Hunters*, edited by Kevin J. Anderson. Reading these books late at night in Ottawa, Ontario, my mind began to imagine what life out among the stars might actually be like. In a digital age with book reading in decline, it feels vital to remember that books were the original spaceflight technology.

This book is the result of an unusual mix of interests: an interest in spaceflight stories as an input to the production of spaceflight; an interest in the ideas that the stories contain related to development of space expansionism as a belief system; an interest in the spaceflight stories themselves as works of literature; and an interest in the individuals who wrote these stories. I suspect the work may not have been overly satisfactory to anyone looking for a comprehensive assessment of any single one of these interests across the works covered; however, I hope that others with a similar or related set of interests may nonetheless find something of value in this eclectic book.

As a work that began as a sort of "lost chapter" of *The Long Space Age*, I want to offer my profoundest thanks again to everyone who supported and was part of that earlier work, and who thereby also helped this one. My thanks in particular to my parents, Thomas and Susan MacDonald, who not only encouraged a love of books, but who also supported my early enthusiasm for spaceflight narratives and helpfully took me to see two vital spaceflight narratives—*Return of the Jedi* and *Spaceballs*—during their original theatrical releases.

My thanks to my editor, Susan Buckley, at the MIT Press, who continued to support this project over the years required to bring it to fruition, as well as to Kathleen Caruso and Melinda Rankin for their excellent editing, proofreading, and fact-checking support. My sincere thanks are given also for the friendship and support of Matthew Daniels and David Lasker, who allowed me the use of their wonderful cabin in Wardensville, West Virginia, to write, and for Nicholas Skytland, who generously provided his container cabin in Otto, North Carolina, during the final editing phase. Warmest thanks also to my Airbnb hosts in Chisinau, Moldova, and to the exemplary staff at the Raffles Dubai, where important phases of writing and editing occurred.

Most importantly, this book would not have happened without the encouragement and support of my partner for many years, Mackenzie Huffman. It was during the COVID-19 pandemic that she suggested that I revisit my interest in early spaceflight fiction and turn that interest into a book. My deepest thanks go to her for all of the spaceflight narratives we shared together—from *The Expanse*, to the International Space Station National Laboratory, to the launch of Tanager-1, to every *Civilization VI* science victory.

Finally, to everyone everywhere who imagines or produces spaceflight narratives, thank you. If it wasn't for you, we would never believe that we could build the machines that enable us to leave Earth and explore the cosmos.

NOTES

INTRODUCTION

1. Konstantin Tsiolkovsky, "Is This Mere Fantasy?," in *The Science Fiction of Konstantin Tsiolkovsky*, ed. Adam Starchild (University Press of the Pacific, 1979), 453–455.

2. NASA, *NASA Economic Impact Report: October 2024* (NASA, 2024).

3. IMBD, "Top Lifetime Grosses," accessed April 17, 2025, https://www.boxofficemojo.com/chart/top_lifetime_gross/. The top-grossing movies with elements of spaceflight narrative are *Star Wars: Episode VII—The Force Awakens*; *Avengers: Endgame*; *Avengers: Infinity War*; *Avatar*; and *Avatar: The Way of Water*.

4. Marjorie Hope Nicolson, *Voyages to the Moon* (Macmillan, 1948); Roger Lancelyn Green, *Into Other Worlds: Space-Flight in Fiction, from Lucian to Lewis* (Abelard-Schuman, 1958); William Sims Bainbridge, *The Spaceflight Revolution: A Sociological Study* (John Wiley & Sons, 1976); George Locke, *Voyages in Space: A Bibliography of Interplanetary Fiction, 1801–1914* (Ferret Fantasy, 1975); Ron Miller, *The Dream Machines: A Pictorial History of the Spaceship in Art, Science and Literature* (Krieger, 1993); Howard E. McCurdy, *Space and the American Imagination* (Smithsonian Books, 1997); and De Witt Douglas Kilgore, *Astrofuturism: Science, Race, and Visions of Utopia in Space* (University of Pennsylvania Press, 2003).

5. Wernher von Braun and Frederick Ordway, *History of Rocketry and Space Travel* (Crowell, 1966); and Willy Ley, *Rockets and Space Travel: The Future of Flight Beyond the Stratosphere* (Viking, 1947).

6. John Rieder, *Colonialism and the Emergence of Science Fiction* (Wesleyan University Press, 2008).

7. Sam Moskowitz, *Explorers of the Infinite: Shapers of Science Fiction* (World Publishing Company, 1963); Brian Wilson Aldriss, *Billion Year Spree: The True History of*

Science Fiction (Doubleday, 1973); Adam Roberts, *The History of Science Fiction*, 2nd ed. (Palgrave Macmillan, 2016).

8. Daniel Deudney, *Dark Skies: Space Expansionism, Planetary Geopolitics, and the Ends of Humanity* (Oxford University Press, 2020), 8. Deudney's definition: "Space expansionism is a subset of technological futurism, which creates 'technological imaginaries' that are elaborate visions of future technological possibilities and their potential consequences. Technological futurism has been prominent facet of Promethean modernism since its beginning. Space futurism, what some call 'astrofuturism', is only a small part of technological futurism."

9. Peter E. Gordon, "What Is Intellectual History? A Frankly Partisan Introduction to a Frequently Misunderstood Field" (Spring 2012), https://ces.fas.harvard.edu/uploads/files/Reports-Articles/What-is-Intellectual-History-Essay-by-Peter-Gordon.pdf.

10. James Pethokoukis, "Space Expansionism, Geopolitics, and the Future of Humanity: My Long-Read Q&A with Daniel Deudney," American Enterprise Institute, June 19, 2021, https://www.aei.org/economics/space-expansionism-geopolitics-and-the-future-of-humanity-my-long-read-qa-with-daniel-deudney/.

11. David Lasser, *The Conquest of Space* (Apogee Books, 2002); Arthur C. Clarke, *The Exploration of Space* (Harper, 1951); Wernher von Braun, *The Mars Project* (University of Illinois Press, 1953); John F. Kennedy, "Address at Rice University on the Nation's Space Effort," September 12, 1962, https://www.rice.edu/jfk-speech; Gerard K. O'Neill, *The High Frontier: Human Colonies in Space* (Space Studies Institute Press, 1989); Carl Sagan, *Pale Blue Dot: A Vision of the Human Future in Space* (Random House, 1994); Blue Origin, "For the Benefit of Earth," presentation at Blue Origin , May, 9, 2019, uploaded May 10, 2019, YouTube, 51 min., 27 sec., https://www.youtube.com/watch?v=GQ98hGUe6FM; and Elon Musk., "The Road to Making Life Multiplanetary," presentation at Starbase, May 29, 2025, uploaded May 30, 2025, YouTube, 40 min., 36 sec., https://www.youtube.com/watch?v=vse4nvYdtQE.

12. For an evocative exploration of how the spaceflight narrative and the idea of an expansive human future in space has impacted modern long-termism and effective altruism, see Adam Becker, *More Everything Forever: AI Overlords, Space Empires, and Silicon Valley's Crusade to Control the Fate of Humanity* (Basic Books, 2025). I would also add that there is a case to be made that long-termism and effective altruism, and their intersections with the spaceflight narrative and space expansionism, might productively be described together as *space existentialism*—and we would benefit from further literary and intellectual histories of these threads as well.

13. Quoted in Lester M. Hirsch, ed., *Man and Space* (Pitman, 1966), 77.

14. Emília Pásztor, "Nebra Disk," in *Handbook of Archaeoastronomy and Ethnoastronomy*, ed. C. Ruggles (Springer, 2015).

15. Alexander MacDonald, *The Long Space Age: The Economic Origins of Space Exploration from Colonial America to the Cold War* (Yale University Press, 2017).

16. For those interested in additional narratives beyond those covered here, see in particular Locke, *Voyages in Space*, and Miller, *Dream Machines*.

CHAPTER 1

1. The most accessible version of this text is found in J. Wilkins, *The Mathematical and Philosophical Works of the Right Reverend John Wilkins* (London, [1640] 1708), 113.

2. For theory and discussion on the processes of technology development and evolution, see W. Brian Arthur, *The Nature of Technology: What It Is and How It Evolves* (Free Press, 2009).

3. Lucian of Samosata, *Lucian's Wonderland: Being a Translation of the "Vera Historia"*, trans. St. John Basil Wynne Willson (London, 1899).

4. Lucian of Samosata, *The Works of Lucian*, vol. 3, trans. Thomas Francklin (London, 1781), 354–389.

5. Firdausi of Tus, *The Shāhnāma of Firdausi*, vol. 2, trans. Arthur George Warner and Edmond Warner (Kegan Paul, Trench, Trübner & Co., 1906), 103–105; Lodovico Ariosto, *Orlando Furioso*, trans. William Stewart Rose (National Alumni, 1906), 295–298; and Francis Godwin, *The Man in the Moone: Or a Discourse of a Voyage Thither* (London, 1638).

6. Johannes Kepler, *Kepler's Somnium: The Dream, or Posthumous Work on Lunar Astronomy*, trans. Edward Rosen (University of Wisconsin Press, 1967), 15–16; and Gale E. Christianson, "Kepler's *Somnium*: Science Fiction and the Renaissance Scientist," *Science Fiction Studies* 3, no. 1 (March 1976): 79–90.

7. Derek Sayer, *Prague: Crossroads of Europe* (Reaktion Books, 2018), 25.

8. Kepler, *Kepler's Somnium*, 15.

9. *Kepler's Somnium*, 15. Kepler's own footnote on his choosing of this name shows a delightful flair for the creative: "'Moon' in Hebrew is Lebana or Levana. I could have called it 'Selenitis' but Hebrew words, being less familiar to our ears, inspire greater awe and are recommended in the occult arts" (Kepler, 53).

10. Kepler, *Kepler's Somnium*,15.

11. Kepler, *Kepler's Somnium*, 16.

12. Kepler, *Kepler's Somnium*, 16.

13. Kepler, *Kepler's Somnium*, 27.

14. Kepler, *Kepler's Somnium*, 33.

15. Johannes Kepler, *Kepler's Conversation with Galileo's Sidereal Messenger*, trans. Edward Rosen (Johnson Reprint Corporation, 1965), 39.

16. Gabrielle Verdier, *Charles Sorel* (Twayne, 1984), 1.

17. Charles Sorel, *The Comical History of Francion* (London, 1655), book 11, 3.

18. Sorel, *Comical History of Francion*, book 11, 5.

19. Godwin, *The Man in the Moone*, 22–27.

20. Godwin, *The Man in the Moone*, 46.

21. Godwin, *The Man in the Moone*, 65.

22. Edgar Allan Poe, *The Works of the Late Edgar Allan Poe* (London, 1857), 49.

23. For more on the context of Godwin and Wilkins, see H. W. Lawton, "Bishop Godwin's *Man in the Moone*," *Review of English Studies* 7, no. 25 (January 1931): 23–55; Allan Chapman, "'A World in the Moon': John Wilkins and His Lunar Voyage of 1640," *Quarterly Journal of the Royal Astronomical Society* 32, no. 2 (1991): 121–132; and David Cressy, "Early Modern Space Travel and the English Man in the Moon," *American Historical Review* 111, no. 4 (October 2006): 961–982.

24. John Wilkins, *The Discovery of a World in the Moone: Or, A Discourse Tending to Prove That 'tis Probable There May Be Another Habitable World in That Planet* (London, 1638).

25. Ben Jonson, "News from the World in the Moon Discovered in the Moon, a Masque," in *The Works of Ben Jonson*, ed. W. Gifford (London, 1879), 615.

26. Wilkins, *Discovery of a World*, 213–215.

27. Wilkins, *Discovery of a World*, A4.

28. Wilkins, *Discovery of a World*, 155.

29. Wilkins, *Discovery of a World*, 160.

30. Thomas P. Harrison, T., "Birds in the Moon," *Isis* 45, no. 4 (December 1954): 323–330.

31. Wilkins, *Discovery of a World*, 159.

32. Wilkins, *Discovery of a World*, 136.

33. Wilkins, *Discovery of a World*, 138.

34. Wilkins, *Discovery of a World*, 159.

35. Wilkins, *Discovery of a World*, 241–242, 180. See also Carl Sagan, *Pale Blue Dot: A Vision of the Human Future in Space* (Random House, 1994), 6–7; note the similarity in theme and style of the first two sentences of chapter 1.

36. John Wilkins, *The Mathematical and Philosophical Works of the Right Rev. John Wilkins* (London, 1802), vol. 1, book 1, 113.

37. Ron Miller, *Dream Machines: A Pictorial History of the Spaceship in Art, Science and Literature* (Krieger, 1993), 9.

38. Chapman, "'World in the Moon,'" 129–130.

39. Savinien de Cyrano de Bergerac, *A Voyage to the Moon*, trans. A. Lovell (New York, 1899), 39.

40. Cyrano de Bergerac, *Voyage to the Moon*, 37–42.

41. Cyrano de Bergerac, *Voyage to the Moon*, 22–32.

42. Margaret Cavendish, *New Blazing World and Other Writings* (New York University Press, 1992).

43. Cavendish, *New Blazing World and Other Writings*, 184–185.

44. A. Behn, *The Emperor of the Moon: A Farce. As it is Acted by Their Majesties Servants, At the Queens Theater* (London, 1687), 4; emphasis in the original.

45. Behn, *Emperor of the Moon*, 4.

46. David Russen, *Iter Lunare: Or, a Voyage to the Moon* (London, 1703), 17–18.

47. Russen, *Iter Lunare*, 32–24.

48. Russen, *Iter Lunare*, 22–28, 29–30.

49. Russen, *Iter Lunare*, 29. It is worth noting that NASA and its international and industry partners had begun to develop a first way station on the way to the Moon, the Gateway, though that program has now been paused at the time of writing. Nonetheless, should a name be needed for a future watering hole on the way to the Moon, one could do worse than Russen's Olde Cislunar Inn.

50. Russen, *Iter Lunare*, 21, 61.

51. Russen, *Iter Lunare*, 18.

52. Russen, *Iter Lunare*, 40.

53. Russen, *Iter Lunare*, 42.

54. Russen, *Iter Lunare*, 42.

55. Russen, *Iter Lunare*, 34–36, 61.

56. Russen, *Iter Lunare*, 44–45.

57. Russen, *Iter Lunare*, 47.

58. Russen, *Iter Lunare*, 49.

59. Russen, *Iter Lunare*, 49–50.

60. Russen, *Iter Lunare*, 62.

61. Russen, *Iter Lunare*, 48.

62. *A New Journey to the World in the Moon* (London, 1741), 5.

63. *New Journey to the World in the Moon*, 5.

64. Samuel Brunt, *A Voyage to Cacklogallinia* (London, 1727), 108–119.

65. Daniel Defoe, *The Consolidator: or, Memoirs of Sundry Transactions from the World in the Moon* (London, 1705), 36.

66. Defoe, *Consolidator*, 16–17.

67. Marjorie Hope Nicolson, *Voyages to the Moon* (Macmillan, 1948), 185–186.

68. Defoe, *Consolidator*, 51.

69. Defoe, *Consolidator*, 61.

70. Defoe, *Consolidator*, 4.

71. Defoe, *Consolidator*, 17; emphasis in the original.

72. Defoe, *Consolidator*, 43–44.

73. Defoe, *Consolidator*, 52.

74. Defoe, *Consolidator*, 54.

75. Murtagh McDermot, *A Trip to the Moon* (London, 1728).

76. McDermot, *Trip to the Moon*, 84.

77. McDermot, *Trip to the Moon*, 84–85.

78. McDermot, *Trip to the Moon*, 86.

79. McDermot, *Trip to the Moon*, 85.

80. Ralph Morris, *A Narrative of the Life and Astonishing Adventures of John Daniel* (London, 1751), 146.

81. Morris, *John Daniel*, 181.

82. Morris, *John Daniel*, 181.

83. Morris, *John Daniel*, 197, 211.

84. Morris, *John Daniel*, 222.

85. Morris, *John Daniel*, 225.

86. Morris, *John Daniel*, 235.

87. Morris, *John Daniel*, 311–312.

88. Voltaire, *Micromégas*, trans. M. Beachot (Paris, 1829).

89. Marie-Anne de Roumier-Robert, *The Voyages of Lord Seaton to the Seven Planets*, trans. Brian Stableford (Black Coat Press, 2015), 7.

90. Roumier-Robert, *Voyages of Lord Seaton*, 78.

91. Roumier-Robert, *Voyages of Lord Seaton*, 25.

92. Louis Guillaume de La Follie, *Le Philosophe sans Prétention, ou L'homme Rare. Ouvrage Physique, Chymique, Politique et Moral* (Paris, 1775), 30–34.

93. Faith K. Pizor and T. Allan Comp, eds., *The Man in the Moone, and Other Lunar Fantasies* (Praeger, 1971), 149.

94. Nicholas Lunatic, *Satiric Tales: Consisting of A Voyage to the Moon; All the Tailors, or The Old Cloak; and the Fat Witch of London* (London, 1808); and Miller, *Dream Machines*, 34–35.

95. Miller, *Dream Machines*, 14

96. Eberhard Christian Kindermann, *The Speedy Journey*, trans. Dwight Decker (CreateSpace, 2014).

97. Kindermann, *Speedy Journey*, 50.

CHAPTER 2

1. Clark Blaise, *Time Lord: Sir Sandford Fleming and the Creation of Standard Time* (Vintage Books, 2000), 178.

2. Daniel Moore, *An Account of Count D'Artois and His Friend's Passage to the Moon* (Litchfield, CT, 1785).

3. Moore remains a very obscure figure, with almost nothing known about him other than his birth in 1764 in Salisbury, Connecticut, and death in 1822 in New

York City. A likely fascinating early American biography awaits the scholar who can uncover more.

4. Moore, *Account of Count D'Artois*, 5.

5. Moore, *Account of Count D'Artois*, 6.

6. Moore, *Account of Count D'Artois*, 11–12.

7. Moore, *Account of Count D'Artois*, 12–13.

8. Alexander MacDonald., *The Long Space Age: The Economic Origins of Space Exploration from Colonial America to the Cold War* (Yale University Press, 2017), 45–46.

9. Moore, *Account of Count D'Artois*, 15.

10. Moore, *Account of Count D'Artois*, 17.

11. Moore, *Account of Count D'Artois*, 19. As a side note, I hope that future NASA astronauts consider honoring this line with perhaps a similar statement prior to a future landing on the lunar surface: "Houston, we shall in a few minutes land in a world of which we are wholly unacquainted; but let it not ruffle us."

12. Moore, *Account of Count D'Artois*, 20.

13. Moore, *Account of Count D'Artois*, 21.

14. Moore, *Account of Count D'Artois*, 27.

15. Moore, *Account of Count D'Artois*, 30.

16. Moore, *Account of Count D'Artois*, 33.

17. Mary-Jane Rubenstein, *Astrotopia: The Dangerous Religion of the Corporate Space Race* (University of Chicago Press, 2022); and Kendrick Oliver, *To Touch the Face of God: The Sacred, the Profane, and the American Space Program, 1957–1975* (Johns Hopkins University Press, 2013).

18. Washington Irving, *A History of New York From the Beginning of the World to the End of the Dutch Dynasty* (New York, 1820).

19. Irving, *History of New York*, 80.

20. Irving, *History of New York*, 81.

21. Geo Fowler, *A Flight to the Moon; or, the Vision of Randalthus* (Baltimore, 1813).

22. Fowler's book is but the first in a long line of American spaceflight fiction that relies on dreams, spirit journeys, or other metaphysical means to transport its characters through the solar system. Because these tales do not have any technical or economic content pertaining to the development of spaceflight concepts, they are not considered here, with the exception of *A Flight to the Moon*. However, the genre does include some influential early works—including, by the early twentieth century, Edgar Rice Burroughs's popular pulp fiction Mars stories, in which the hero, John Carter, reaches Mars by astral projection. For those interested in the nineteenth-century American spaceflight forerunners in this tradition, see Joel R. Peabody, *A World of Wonders* (Boston, 1838); *Experiences of Eon and Eona* (Springfield, MA, 1886); Henry A. Gaston, *Mars Revealed* (San

Francisco, 1880); Spirito Gentil, *Earth-Born!* (New York, 1889); Camille Flammarion, *Urania* (Boston, 1890); C. C. Dail, *Willmoth the Wanderer, or, The Man from Saturn* (New York, 1891); Francis Worcester Doughty, *Mirrikh, or, A Woman from Mars* (New York, 1892); F. M. Clarke, *A Maiden of Mars* (Chicago, 1892); H. M. Bien, *Ben-Beor* (Baltimore, 1891); M. D. Leggett, *A Dream of a Modest Prophet* (Philadelphia, 1890); Mary Platt Parmele, *Ariel* (New York, 1898); and Th. Flournoy, *From India to the Planet Mars* (New York, 1900).

23. Fowler, *Flight to the Moon*, 6.

24. Fowler, *Flight to the Moon*, 30–31.

25. Fowler, *Flight to the Moon*, 31.

26. Fowler, *Flight to the Moon*, 32.

27. Fowler, *Flight to the Moon*, 34.

28. Fowler, *Flight to the Moon*, 46.

29. James Fieser, ed., *The Life and Philosophy of George Tucker* (Thoemmes Continuum, 2004), vii.

30. George Tucker, *A Voyage to the Moon: With some Account of the Manners and Customs, Science and Philosophy, of the People of Morosofia, and Other Lunarians* (New York, 1827), 12–14, 14–19.

31. Francis Godwin, *A voyage to the moon: with an account of the religion, laws, customs, and manner of government among the lunars or moon-men: also a description of the country and the palaces of their princes, together with their language and the ceremonies us'd at the funerals of their deceas'd* (Stamford, 1718).

32. Anonymous, "Voyage to the Moon," *American Quarterly Review*, vol. 3 (1828): 61–88. Note, however, that authorship of the anonymous review was later claimed by Tucker's close friend and colleague Robley Dunglison, professor of anatomy and medicine at the University of Virginia. See "Obituary Notice of Prof. George Tucker," *Proceedings of the American Philosophical Society* 9, no. 68 (1862): 64–70, 65.

33. Tucker, *Voyage to the Moon*, 30.

34. Tucker, *Voyage to the Moon*, 35.

35. Tucker, *Voyage to the Moon*, 23.

36. Tucker, *Voyage to the Moon*, 45.

37. Tucker, *Voyage to the Moon*, 21.

38. Tucker, *Voyage to the Moon*, 43.

39. Tucker, *Voyage to the Moon*, 49.

40. Anonymous, "Voyage to the Moon," 70.

41. Tucker, *Voyage to the Moon*, 42.

42. Tucker, *Voyage to the Moon*, 30, 43.

43. Tucker, *Voyage to the Moon*, 215.

44. Tucker, *Voyage to the Moon*, 41.

45. Ron Miller, *Dream Machines: A Pictorial History of the Spaceship in Art, Science and Literature* (Krieger, 1993), 29.

46. Arthur Hobson Quinn, *Edgar Allan Poe: A Critical Biography* (John Hopkins University Press, 1998), 539.

47. Edgar Allan Poe, "The Unparalleled Adventure of One Hans Pfaall," in *The Works of the Late Edgar Allan Poe* (New York, 1850), 51.

48. For Poe's context in the literary tradition of lunar speculations, see Maurice J. Bennet, "Edgar Allan Poe and the Literary Tradition of Lunar Speculation," *Science Fiction Studies* 10, no. 2 (1983): 137–147. For a full discussion of Poe's sources for "Hans Pfaall," see William H. Gravely, "A Note on the Composition of Poe's 'Hans Pfaall,'" *Poe Newsletter* 3, no. 1 (1970): 2–5; R. Wilkinson, "Poe's 'Hans Pfaall' Reconsidered," *Notes and Queries*, 13, no. 9 (1966): 333–337; J. O. Bailey, "Sources for Poe's *Arthur Gordon Pym*, 'Hans Pfaall,' and Other Pieces," *PMLA* 57, no. 2 (1942): 513–535; and Meredith Neill Posey, "Notes on Poe's 'Hans Pfaall,'" *Modern Language Notes* 45, no. 8 (1930): 501–507.

49. Poe, "Unparalleled Adventure of One Hans Pfaall."

50. Edgar Allan Poe, "The Literati [Part VI]: Richard Adams Locke," in *Godey's Lady's Book*, October 1846, no. 33, 159–161.

51. Poe, "Unparalleled Adventure of One Hans Pfaall," 15–18; and N. A. Rynin, *Dreams, Legends, and Early Fantasies*, vol. 1 of *Interplanetary Flight and Communications* (Israel Program for Scientific Translations, 1970), 94.

52. Poe, "Unparalleled Adventures of One Hans Pfaall," 51.

53. This concept was still taught in the early twentieth century; see Poe, "Unparalleled Adventures of One Hans Pfaall," 17–18; John F. W. Herschel, *A Treatise on Astronomy* (Philadelphia, 1851), 293; M. Arago, *Tract on Comets; and Particularly on the Comet That Is to Intersect the Earth's Path in October, 1832*, trans. John Farrar (Boston, 1832), 44–45; and Simon Newcomb, *Popular Astronomy* (New York, 1899), 393–395.

54. Poe, "Unparalleled Adventures of One Hans Pfaall," 17–18.

55. Poe, "Hans Pfaall," 38–41; Roderick W. Home, "The Origin of the Lunar Craters: An Eighteenth-Century View," *Journal for the History of Astronomy* 3, no. 1 (1972): 1–10; and William Graves Hoyt, "G. K. Gilbert's Contribution to Selenology," *Journal for the History of Astronomy* 13, no. 3 (1982): 155–167.

56. Johann Heinrich von Mädler's *Mappa Selenographica* and *Der Mond* in the 1830s convinced most scientists that there was no lunar atmosphere, but the concept was still taught in school well into the 1850s. See Poe, "Unparalleled Adventures of One Hans Pfaall," 41–42; and John Brocklesby, *Elements of Astronomy, for Schools and Academies, with Explanatory Notes, and Questions for Examination* (New York, 1855), 160–161.

57. Poe, "Unparalleled Adventures of One Hans Pfaall," 7.

58. Poe, "Hans Pfaall," 7.

59. Poe, "Hans Pfaall," 7.

60. Poe, "Hans Pfaall," 8–9.

61. Not only was Poe the subject of Verne's only literary review, but Verne also has his space explorers in *From the Earth to the Moon* refer to and cheer Poe and Hans Pfaall, and he also set the Gun Club—the organization that undertook the journey to the Moon—in Baltimore, the city in which Poe spent most of his life, likely as a tribute to Poe. Harold Beaver, "Introduction," in *The Science Fiction of Edgar Allan Poe* (Penguin, 1976), ix; and Jules Verne, "Edgar Poe et ses Ouevres," *Musée des Familles* 31, no. 7 (1864): 193–208.

62. The spaceflight potential of the balloon remained a popular theme in America throughout the nineteenth century, with the short story "To Venus by Balloon" appearing in the first issue of *Frank Leslie's Boys' and Girls' Weekly* in 1866 (see Miller, *Dream Machines*, 54), and an entire balloon-borne tour of the solar system appearing as late as 1873 in Paul Aermont [pseud.], *Narrative of the Travels and Adventures of Paul Aermont Among the Planets* (Boston, 1873). There was also a facetious "ego-as-balloon" voyage to the Moon written before Riddell's tale; see Aerio-Nautical Man [pseud.], "Recollections of Six Days' Journey in the Moon: Part I," *Southern Literary Messenger*, July 1844, 434–437; and Aerio-Nautical Man [pseud.], "Recollections of Six Days' Journey in the Moon: Part II," *Southern Literary Messenger*, August 1844, 492–495. The two-part story is a satire in the traditional vein that nonetheless provides an interesting discussion of the aerio-nautical man's motivations for spaceflight: "This terrestrial world of ours has lately been so thoroughly explored by all sorts of ingenious and inquisitive traveler, who have left nothing for those that may come after them either to describe or invent." Here, in the 1840s, with the Oregon trail still in its infancy and one year before President Polk's invocation of America's "manifest destiny" in support of the Monroe Doctrine, is an early expression of that seemingly timeless pro-spaceflight argument: that there is nowhere left to explore on Earth and that humanity must thus look to the heavens for new adventures. Leaving aside the question of the appropriateness of this rhetoric—then and now—the sentiment nonetheless continues to present voyages into space as a personal act, in this case drawing specifically on private, intrinsic motivations.

63. The most recent account of his career is Everett F. Bleiler, "John Leonard Riddell, Pioneer," *Science Fiction Studies* 36, no. 2 (2009): 284–299. See also Ralph W. Dexter, "The Early Career of John L. Riddell as a Science Lecturer in the 19th Century," *Ohio Journal of Science* 88, no. 5 (1988): 184–188; Hubert Skinner and Karlem Riess, "John Leonard Riddell: From Rensselaer to New Orleans (1827–1865)," *Earth Sciences History* 4, no. 1 (1985): 75–80; Karlem Riess, *John Leonard Riddell: Scientist-Inventor, Melter and Refiner of the New Orleans Mint, 1839–1848, Postmaster of New Orleans, 1859–1862* (Louisiana Heritage Press, 1977); James O. Breeden, ed., *A Long Ride in Texas: The Explorations of John Leonard Riddell* (Texas A&M University Press, 1994); and L. H. Bailey Jr., "Some North American Botanists: John Leonard Riddell," *Botanical Gazette* 8, no. 8 (1883): 269–271. There is also a brief mention of Riddell in Thomas D. Crouch, "'To Fly to the World in the Moon': Cosmic Voyaging in Fact and Fiction from Lucian to Sputnik," in *Science Fiction and Space Futures: Past and Present*, ed. Eugene M. Emme (Univelt, 1982), 14.

64. United States Congress, "Report No. 8: A.P. Field Report," in *Reports of Committees of the House of Representatives Made During the First Session Thirty-Eighth Congress, 1863–1864* (Washington, DC, 1864). According to the congressional report on the

matter, Riddell claims to have been elected governor by five or six hundred people in New Orleans, during a time when voting for a governor and congressional representatives had been forbidden by the then-military governor of the state. The report states that Riddell claimed to have taken the oath of office before a justice of the peace in New Orleans and begun his term in January of 1864, a year before his death. As the military governorship of Brigadier General George Foster Shepley held the de facto power in the state at the time and had suppressed voting in New Orleans, Riddell's "governorship" was limited in its significance.

65. Riess, *John Leonard Riddell*, 45.

66. J. L. Riddell, *Orrin Lindsay's Plan of Aerial Navigation, with a Narrative of His Explorations in the Higher Regions of the Atmosphere, and His Wonderful Voyage Round the Moon!* (New Orleans, 1847), 13.

67. Riddell, *Plan of Aerial Navigation*, 5–6.

68. Riddell, *Plan of Aerial Navigation*, 6.

69. Riddell, *Plan of Aerial Navigation*, 9–10.

70. Riddell, *Plan of Aerial Navigation*, 11.

71. Riddell, *Plan of Aerial Navigation*, 13–14.

72. Riddell, *Plan of Aerial Navigation*, 7–8.

73. Riddell, *Plan of Aerial Navigation*, 14.

74. Riddell, *Plan of Aerial Navigation*, 14.

75. Riddell, *Plan of Aerial Navigation*, 15.

76. Riddell, *Plan of Aerial Navigation*, 16.

77. Riddell, *Plan of Aerial Navigation*, 20–21.

78. Riddell, *Plan of Aerial Navigation*, 22.

79. Riddell, *Plan of Aerial Navigation*, 24

80. Riddell, *Plan of Aerial Navigation*, 25.

81. Riddell, *Plan of Aerial Navigation*, 26.

82. Riddell, *Plan of Aerial Navigation*, 26–27.

83. Riddell, *Plan of Aerial Navigation*, 27.

84. Riddell, *Plan of Aerial Navigation*, 27.

85. Riddell, *Plan of Aerial Navigation*, 28.

86. Riddell, *Plan of Aerial Navigation*, 32.

87. Riddell, *Plan of Aerial Navigation*, 32.

88. Riddell, *Plan of Aerial Navigation*, 33.

89. Noel Deisch Papers, 1910–1937, Library of Congress; "Would Take Man Year to Reach Moon in Rocket," *Brooklyn Daily Eagle* , June 30, 1929, 14.

90. Elbert Perce, *Gulliver Joi: His Three Voyages; Being an Account of His Marvelous Adventures in Kailoo, Hydrogenia and Ejario* (New York, 1852); and Miller, *Dream Machines*, 43.

91. "Obituary: Mr. Elbert Perce," *New York Times*, January 21, 1869; and *The American Annual Cyclopaeida and Register of Important Events of the Year 1869*, vol. 9 (New York, 1870), 496.

92. "Obituary: Mr. Elbert Perce."

93. "Obituary: Mr. Elbert Perce"; advertisement for Perce's Magnetic Globes, *Illinois Teacher: Devoted to Education, Science, and Free Schools*, vol. 11 (1865): 103–104.

94. "Obituary: Mr. Elbert Perce."

95. Perce, *Gulliver Joi*, 30–31.

96. Perce, *Gulliver Joi*, 33.

97. Perce, *Gulliver Joi*, 33.

98. Perce, *Gulliver Joi*, 34.

99. Perce, *Gulliver Joi*, 35.

100. Thomas Kettell, ed., *The United States Magazine and Democratic Review*, vol. 29 (New York, 1851).

101. Perce, *Gulliver Joi*, 271.

102. Perce, *Gulliver Joi*, 272.

103. *Pathways to Exploration: Rationales and Approaches for a U.S. Program of Human Space Exploration* (National Academies Press, 2014).

CHAPTER 3

1. Marjorie Hope Nicholson, *Voyages to the Moon* (Macmillan, 1948), 243.

2. Ron Miller, "How Jules Verne Invented Astronautics," *Gizmodo*, May 24, 2014, https://gizmodo.com/how-jules-verne-invented-astronautics-1493029901.

3. Walter James Miller, ed., *The Annotated Jules Verne: From the Earth to the Moon* (Thomas Y. Crowell, 1978); Jules Verne, *From the Earth to the Moon: Annotated for our Spacefaring Age*, ed. Anastasia Klimchynskaya (MIT Press, 2026).

4. Jean Jules-Verne, *Jules Verne: A Biography* (Taplinger, 1976), 5–6.

5. Jules-Verne, *Jules Verne: A Biography*, 57–58.

6. William Butcher, *Jules Verne: The Definitive Biography* (Thunder's Mouth Press, 2006), 225.

7. Jules Verne, "Edgar Poe et ses Ouevres," *Musée des Familles* 31, no. 7 (1864): 193–208.

8. Harold Beaver, "Introduction," in *The Science Fiction of Edgar Allan Poe* (Penguin, 1976), ix; Verne, "Edgar Poe et ses Ouevres."

9. Butcher, *Jules Verne*, 165.

10. Jules-Verne, *Jules Verne*, 93.

11. Ron Miller, "Spaceflight and Popular Culture," in *Societal Impact of Spaceflight*, ed. Steven J. Dick and Roger D. Launius (National Aeronautics and Space Administration, Office of External Relations, History Division, 2007), 506.

12. Jacques Crovisier, "Astronomy and Astronomers in Jules Verne's Novels," in *The Role of Astronomy in Society and Culture: Proceedings IAU Symposium No. 260*, ed. D. Valle-Gabaud and A. Boksenberg (International Astronomical Union, 2009), 322.

13. Jules Verne, *From the Earth to the Moon, Direct in Ninety-Seven Hours and Twenty-Minutes: And a Trip Round It*, trans. Louis Mercier and Eleanor E. King (New York, 1874), 20; emphasis in the original.

14. Mission Evaluation Team, *Apollo 11 Mission Report*, MSC-00171 (Manned Spacecraft Center, NASA, 1969).

15. Verne, *From the Earth to the Moon*, 22.

16. Gerald V. Bull, "Development of Gun Launched Vertical Probes for Upper Atmosphere Studies," *Canadian Aeronautics and Space Journal* 10 (1964): 236–247; and Dale Grant, *Wilderness of Mirrors: The Life of Gerald Bull* (Prentice-Hall, 1991). Notably, Gerald Bull's final design for his Project Babylon gun to reach orbit had a barrel length of 156 meters.

17. Tim Fernholtz, "Longshot Is Gunning for Kinetic Space Launch," *Payload*, June 23, 2025, https://payloadspace.com/longshot-is-gunning-for-kinetic-space-launch/.

18. Verne, *From the Earth to the Moon*, 105.

19. Verne, *From the Earth to the Moon*, 38.

20. Verne, *From the Earth to the Moon*, 11–12.

21. Verne, *From the Earth to the Moon*, 12–13.

22. Verne, *From the Earth to the Moon*, 13.

23. Verne, *From the Earth to the Moon*, 13.

24. Nicholson, *Voyages to the Moon*, 243.

25. Miller, *Annotated Jules Verne*, 66–71.

26. Calculations performed using inflation adjustments at MeasuringWorth (https://measuringworth.com).

27. Linda Neuman Ezell, *NASA Historical Data Book: Volume III: Programs and Projects 1969–1978* (NASA, 1988), 61.

28. Jack Speer, *Fancestral Voices* (NESFA Press, 2004); and Frank H. Winter, *Prelude to the Space Age: The Rocket Societies: 1924–1940* (Smithsonian Institution Press, 1983), 74.

29. Verne, *From the Earth to the Moon*, 11.

30. Verne, *From the Earth to the Moon*, 59–60.

31. Miller, *Annotated Jules Verne*, 1–2.

32. Miller, *Annotated Jules Verne*, 3.

33. Verne, *From the Earth to the Moon*, 43.

34. Winter, *Prelude to the Space Age*, 73.

35. Miller, *Annotated Jules Verne*, 69.

36. Miller, *Annotated Jules Verne*, 69.

37. Miller, *Annotated Jules Verne*, 70.

38. Miller, *Annotated Jules Verne*, 71.

39. Verne, *From the Earth to the Moon*, 28, 17, 32.

40. Miller, *Annotated Jules Verne*, 148–149.

41. Miller, *Annotated Jules Verne*, 126; and Henry C. Dethloff, *Suddenly, Tomorrow Came . . . A History of the Johnson Space Center* (NASA, 1993), 122.

42. Miller, *Annotated Jules Verne*, 72.

43. Miller, *Annotated Jules Verne*, 71.

44. Miller, *Annotated Jules Verne*, 62.

45. United States Census Office, *A Compendium of the Ninth Census: June 1, 1870* (Government Printing Office, 1872), 136; and Miller, *Annotated Jules Verne*, 91.

46. Miller, *Annotated Jules Verne*, 92.

47. Jules Verne, *Round the Moon* (Ward, Lock and Co., 1956), 66.

48. Verne, *Round the Moon*, 69–70, 72.

49. Verne, *Round the Moon*, 190.

50. Year of first publication of *Around the Moon*, 1870, taken as year of calculation using MeasuringWorth (https://measuringworth.com).

51. See N. A. Rynin, *Dreams, Legends, and Early Fantasies*, vol. 1 of *Interplanetary Flight and Communications* (Israel Program for Scientific Translations, 1970); George Locke, *Voyages in Space: A Bibliography of Interplanetary Fiction 1801–1914* (Ferret Fantasy, 1975); and Ron Miller, *The Dream Machines: A Pictorial History of the Spaceship in Art, Science and Literature* (Krieger, 1993).

52. Simon Newcomb, *His Wisdom, the Defender: A Story* (Harper & Brothers, 1900).

53. Tom D. Crouch, *Aiming for the Stars: The Dreamers and Doers of the Space Age* (Smithsonian Institution Scholarly Press, 2000), 24–25.

54. Wernher von Braun and Frederick Ordway, *History of Rocketry and Space Travel* (Crowell, 1966), 56.

55. M. Harwit and F. Winter, NASM.1999.00038 Peenemunde Interviews Project: Hermann Oberth, November 14–15, 1987.

56. Milton Lehman, *This High Man: The Life of Robert H. Goddard* (Farrar, Straus, 1963), 138.

57. Quoted in Asif A. Siddiqi, "Jules Verne and the Soviet Cosmic Culture," in Jules Verne, *From the Earth to the Moon: Annotated for Our Spacefaring Age*, ed. Anastasia Klimchynskaya, trans. Walter James Miller (MIT Press, 2026), 292. *From the Earth to the Moon* was translated as *From the Gun to the Moon* for Russian audiences.

58. Jules-Verne, *Jules Verne: A Biography*, 58.

59. Jules Verne, "In the Year 2889," *The Forum* 6, no. 6 (1889): 662.

60. Verne, "In the Year 2889," 664.

61. Verne, "In the Year 2889," 665.

62. Verne, "In the Year 2889," 667.

63. Verne, "In the Year 2889," 668–669.

64. Robert Cromie, *A Plunge into Space* (New York, 1891), 5.

65. Cromie, *Plunge into Space,* 23.

66. Cromie, *Plunge into Space,* 227.

67. Jules Verne, *The Chase of the Golden Meteor* (Bison Books, 1998), 100.

68. Verne, *Chase of the Golden Meteor,* 139.

69. Jerrold Siegel, *Marx's Fate: The Shape of a Life* (Penn State University Press, 1993), 281.

CHAPTER 4

1. Edward Everett Hale, *The Brick Moon and Other Stories* (Boston, 1899), vi.

2. For more on Hale's life and works, see Edward E. Hale Jr., *The Life and Letters of Edward Everett Hale,* vols. 1 and 2 (Little, Brown, and Company, 1917); and Jean Holloway, *Edward Everett Hale: A Biography* (University of Texas Press, 1956). To appreciate his status as a leading American role model of the nineteenth century, see also Mary H. Wade, *Real Americans* (Little, Brown, and Company, 1922); and Edwin Wildman, *Famous Leaders of Character: In America from the Latter Half of Nineteenth Century: The Life Stories of Boys Who Have Impressed Their Personalities on the Life and History of the United States* (Page Company, 1922). Within space history literature, the description of Hale in *Exploring the Unknown Vol. 1* contains a number of inaccuracies. It was his uncle—Edward Everett—that gave the two-hour Gettysburg Address before Lincoln's iconic two-minute speech, not Hale himself. There is also no mention of the notion that the "The Brick Moon" was inspired by Richard Adams Locke's "Great Moon Hoax" in the introduction to the 1899 edition of "The Brick Moon"—which Sam Moskowitz initially cites in support for this idea—though Hale evidently did read Locke's work. Hale's supposed quote on the origin of the story is also a misquote from *The Brick Moon and Other Stories,* where his statement appears.

3. "The Rev. Edward E. Hale. His Methods of Work—The Story of the Man Without a Country," *New York Times,* May 6, 1880.

4. Edward Everett Hale, *The Works of Edward Everett Hale,* vol. 4 (Boston, 1899), 3.

5. To this day, the indispensable biography of Hale is Holloway, *Edward Everett Hale: A Biography.*

6. Holloway, *Edward Everett Hale: A Biography,* 3.

7. Holloway, *Edward Everett Hale: A Biography,* 5.

8. Wildman, *Famous Leaders of Character,* 85; Wade, *Real Americans,* 252; and Holloway, *Edward Everett Hale: A Biography,* 13.

9. Holloway, *Edward Everett Hale: A Biography,* 20.

10. Holloway, *Edward Everett Hale: A Biography,* 29, 42.

11. Hale, *Works of Edward Everett Hale,* 3.

12. Hale, *Life and Letters of Edward Everett Hale*, vol. 2, 63.

13. Edward Everett Hale, "The Brick Moon," part 1, *Atlantic Monthly*, October 1869, 451–460, 451.

14. Hale, "Brick Moon," part 1, 453.

15. Hale, "Brick Moon," 454.

16. Hale, "Brick Moon," 456. Orcutt's railway supposedly rose to popularity when he advertised that infant children would travel at three times the normal price, establishing that he was a unique phenomenon: "a railroad manager caring for the comfort of his passengers."

17. Hale, "Brick Moon," 456.

18. Hale, "Brick Moon," 458.

19. Hale, "Brick Moon," 458.

20. Hale, "Brick Moon," 458.

21. Hale, "Brick Moon," 458.

22. Hale, "Brick Moon," 458–459.

23. Hale, "Brick Moon," 459.

24. Hale, "Brick Moon," 459–460.

25. Hale, "Brick Moon," 460.

26. Alexander MacDonald, *The Long Space Age: The Economic Origins of Space Exploration from Colonial America to the Cold War* (Yale University Press, 2017).

27. Eli Thayer, *A History of the Kansas Crusade: Its Friends and Its Foes* (New York, 1899), 125.

28. Holloway, *Edward Everett Hale: A Biography*, 107. The mid-year of the company, 1849, was used as the base year for the calculations.

29. Leverett Wilson Spring, *Kansas: The Prelude to the War for the Union* (Boston, 1894), 32.

30. Thayer, *History of the Kansas Crusade*, xi.

31. Edward Everett Hale, "The Brick Moon," part 2, *Atlantic Monthly*, November 1869, 603–611, 605.

32. Holloway, *Edward Everett Hale: A Biography*, 143–149. In an amazing coincidence, Hale was the civilian aide to the same General Butler who was the military governor of Louisiana, and the object of civilian protest, during fellow spaceflight-imaginer John Leonard Riddell's election as rump governor of the state.

33. Hale, "Brick Moon," part 2, 607.

34. Edward Everett Hale, "The Brick Moon," part 3, *Atlantic Monthly*, December 1869, 679–688, 681.

35. Hale, "Brick Moon," part 3, 682.

36. Edward Everett Hale, "Life in the Brick Moon," *Atlantic Monthly*, February 1870, 215–222.

37. Hale, *Life and Letters of Edward Everett Hale*, vol. 2, 60.

38. Holloway, *Edward Everett Hale: A Biography*, 172.

39. Hale, "Brick Moon," part 3, 683.

40. Hale, "Life in the Brick Moon," 687.

41. Hale, "Life in the Brick Moon," 215–216.

42. Hale, "Life in the Brick Moon," 216.

43. Hale, "Life in the Brick Moon," 220.

44. Hale, "Life in the Brick Moon," 221.

45. Hale, "Life in the Brick Moon," 221.

46. Hale, "Life in the Brick Moon," 219.

47. Hale, "Life in the Brick Moon," 216.

48. Hale, "Life in the Brick Moon," 221.

49. Gerard K. O'Neill, *The High Frontier: Human Colonies in Space* (Morrow, 1976).

CHAPTER 5

1. Florence Carpenter Dieudonné, *Rondah; or, Thirty-Three Years in a Star* (Philadelphia, 1887), 28–29.

2. Robert Godwin, *William Leitch: Presbyterian Scientist & the Concept of Rocket Space Flight 1854–1864* (Apogee Prime, 2017).

3. William Leitch, *God's Glory in the Heavens* (London, 1867), 4.

4. Godwin, *William Leitch: Presbyterian Scientist*, 83.

5. Godwin, *William Leitch: Presbyterian Scientist*, 119.

6. Ives, *Princess of the Moon*, 11.

7. Ives, *Princess of the Moon*, 12.

8. Ives, *Princess of the Moon*, 17.

9. Ives, *Princess of the Moon*, 18.

10. Ives, *Princess of the Moon*, 59.

11. Ives, *Princess of the Moon*, 61–62.

12. Ives, *Princess of the Moon*, 64–65.

13. Ives, *Princess of the Moon*, 67.

14. Ives, *Princess of the Moon*, 70.

15. Ives, *Princess of the Moon*, 72.

16. Washington Gladden, "A Christmas Dinner with the Man in the Moon," in *Santa Claus on a Lark and Other Christmas Stories* (New York, 1890); and Ron Miller, *The Dream Machines: A Pictorial History of the Spaceship in Art, Science and Literature* (Krieger, 1993), 61–63.

17. Polly Cabell, "A Trip to the Moon," *Electra: A Belles Lettres Monthly for Young People*, February 1884, 615–616.

18. Cabell, "Trip to the Moon," 615.

19. See George Locke, *Voyages in Space: A Bibliography of Interplanetary Fiction 1801–1914* (Ferret Fantasy, 1975), 68, for a statistical appreciation of the increasing rate of American interplanetary fiction.

20. Pruning Knife [pseud.], *The Key of Industrial Co-operative Government* (St. Louis, 1886), 11; David T. Burbank, *Reign of the Rabble: The St. Louis General Strike of 1877* (Augustus M. Kelley, 1966), 194; and Charles J. Rooney Jr., *Dreams and Visions: A Study of American Utopias, 1865–1917* (Praeger, 1985), 181.

21. Hudor Genone, *Bellona's Husband: A Romance* (Philadelphia, 1887); "William James Roe," in *Fifty-Third Annual Report of the Association of Graduates of the United States Military Academy* (Seemann & Peters, 1922), 121–128.

22. Genone, *Bellona's Husband*, 25–27.

23. Genone, *Bellona's Husband*, 25.

24. Genone, *Bellona's Husband*, 26.

25. Genone, Bellona's Husband, 26.

26. Edward Bellamy, *Looking Backward: 2007–1887* (Houghton Mifflin, 1917), 255.

27. Anonymous, *Man Abroad: A Yarn of Some Other Century* (Gregg Press, 1976).

28. Kevin H. O'Rourke and Jeffrey G. Williamson, *Globalization and History: The Evolution of a Nineteenth-Century Atlantic Economy* (MIT Press, 1999).

29. Anonymous, *Man Abroad*, 8.

30. Anonymous, *Man Abroad*, 18.

31. Ron Miller, "The Spaceship as Icon: Designs from Verne to the Early 1950s," in *Blueprint for Space*, ed. Frederick Ira Ordway, Randy Liebermann, and Ben Bova (Smithsonian, 1992).

32. Anonymous, *Man Abroad*, 9.

33. Anonymous, *Man Abroad*, 9.

34. Anonymous, *Man Abroad*, 11.

35. Anonymous, *Man Abroad*, 11.

36. Anonymous, *Man Abroad*, 11.

37. Will Stewart, "Collision Orbit," *Astounding Science Fiction*, July 1942, 80–117.

38. Anonymous, *Man Abroad*, 11.

39. Anonymous, *Man Abroad*, 39.

40. Anonymous, *Man Abroad*, 40.

41. Henry George, *The Complete Works of Henry George*, vol. 1, *Progress and Poverty: An Inquiry into the Cause of Industrial Depressions and of Increase of Want with Increase of Wealth: The Remedy* (Doubleday, 1904), 241.

42. Anonymous, *Man Abroad*, 40.

43. Anonymous, *Man Abroad*, 40–41.

44. Anonymous, *Man Abroad*, 41–42.

45. Anonymous, *Man Abroad*, 44.

46. Anonymous, *Man Abroad*, 41.

47. Anonymous, *Man Abroad*, 17–18.

48. Anonymous, *Man Abroad*, 3.

49. Dieudonné, *Rondah.*

50. Frances E. Willard and Mary A. Livermore, eds., *A Woman of the Century: Fourteen Hundred-Seventy Biographical Sketches Accompanied by Portraits of Leading American Women in All Walks of Life* (Buffalo, 1893).

51. Dieudonné, *Rondah*, 27.

52. Dieudonné, *Rondah*, 28.

53. Dieudonné, *Rondah*, 28.

54. Dieudonné, *Rondah*, 28–29.

55. John J. Pierce, *Foundations of Science Fiction: A Study in Imagination and Evolution* (Greenwood Press, 1987), 59.

56. Dieudonné, *Rondah*, 42–43.

57. Dieudonné, *Rondah*, 43.

58. Dieudonné, *Rondah*, 46.

59. Dieudonné, *Rondah*, 58.

60. Dieudonné, *Rondah*, 75.

61. Dieudonné, *Rondah*, 76.

62. Dieudonné, *Rondah*, 77.

63. Dieudonné, *Rondah*, 84.

64. Dieudonné, *Rondah*, 86.

65. Dieudonné, *Rondah*, 87.

66. Dieudonné, *Rondah*, 213–214.

67. Dieudonné, *Rondah*, 222.

68. Dieudonné, *Rondah*, 229.

69. Dieudonné, *Rondah*, 230.

70. William Simpson, *The Man from Mars: His Morals, Politics and Religion* (San Francisco, 1891).

71. Alice Ilgenfritz Jones and Ella Merchant, *Unveiling a Parallel: A Romance* (Boston, 1893).

72. Because the two love stories are sequential, *Unveiling a Parallel* cannot properly be credited with presenting the first interplanetary love or extraterrestrial love

triangle, a scenario that would await two other publications to be first explored in text: Edgar Rice Burroughs, *The Gods of Mars* (Grosset and Dunlap, 1918); and David Lindsay, *A Voyage to Arcturus* (Methuen, 1920).

73. Jones and Merchant, *Unveiling a Parallel*, 1–2.

74. Gustavus W. Pope, *Journey to Mars: the Wonderful World: Its Beauty and Splendor; Its Mighty Races and Kingdoms; Its Final Doom* (New York, 1894); and Gustavus W. Pope, *Journey to Venus: The Primeval World; Its Wonderful Creations and Gigantic Monsters* (Boston, 1895).

75. Henry Olerich, *A Cityless and Countryless World: An Outline of Practical Co-operative Individualism* (Holstein, IA, 1893).

76. Olerich, *Cityless and Countryless World*, 20.

77. Olerich, *Cityless and Countryless World*, 23.

78. Lucy Kavaler, *The Astors: A Family Chronicle of Pomp and Power* (Dodd, Mead & Company, 1966), 119.

79. Kavaler, *The Astors*, 150.

80. Kavaler, *The Astors*, 126, 128.

81. Kavaler, *The Astors*, 141.

82. "Appraisement of Estate Reveals Astor's Personality," *New York Times*, June 22, 1913.

83. "Appraisement of Estate." Because the distribution of these investments over his lifetime is unclear, his mid-life year of 1888 is used.

84. "Appraisement of Estate." At the time of Astor's death in 1912, $1.50 was equivalent to $50.10 in 2024 CPI ratio terms.

85. Incidentally, *Across the Zodiac* is the first story with an interplanetary, extraterrestrial romance, albeit one that is not nearly as intricate or interesting as the two romances in *Unveiling a Parallel*.

86. "John Jacob Astor, Author," *New York Times*, April 15, 1894; and Lewis Randolph Hamersly, *First Citizens of the Republic: An Historical Work* (L. R. Hamersly & Co., 1906), 169–170.

87. John Jacob Astor, *A Journey in Other Worlds: A Romance of the Future* (New York, 1894), 34–35.

88. Astor, *Journey in Other Worlds*, 35, 49–51, 62–66.

89. Astor, *Journey in Other Worlds*, 43–44, 72.

90. Astor, *Journey in Other Worlds*, 37.

91. Astor, *Journey in Other Worlds*, 75.

92. Astor, *Journey in Other Worlds*, 77–78.

93. "Appraisement of Estate." These amounts would be equivalent to $54.8 and $27.4 million in 2023 real price equivalent terms, $425 million and $212.5 million in 2023 PWC ratio equivalent terms, and $2.85 billion and $1.925 billion in 2023 GDP ratio equivalent terms.

94. Astor, *Journey in Other Worlds*, 84.

95. Astor, *Journey in Other Worlds*, 89–90.

96. Astor, *Journey in Other Worlds*, 98–99.

97. Astor, *Journey in Other Worlds*, 99.

98. Astor, *Journey in Other Worlds*, 86.

99. Astor, *Journey in Other Worlds*, 90–91.

100. Astor, *Journey in Other Worlds*, 90–91.

101. Astor, *Journey in Other Worlds*, 100.

102. Kavaler, *The Astors*, 141.

103. Albert Shaw, ed., "A Journey in Other Worlds," in *The Review of Reviews*, vol. 9, January–June 1894 (New York), 753. For a more critical and amusing review from Boston, see "A Journey in Other Worlds," in *The Literary World: A Fortnightly Review of Current Literature*, vol. 25, no. 10 (Boston, 1894), 149.

104. Astor, *Journey in Other Worlds*, 79.

105. Konstantin Tsiolkovsky, *The Science Fiction of Konstantin Tsiolkovsky*, ed. Adam Starchild (University Press of the Pacific, 1979).

CHAPTER 6

1. H. G. Wells, *The War of the Worlds* (London, 1898), 314.

2. Personally, I doubt H. G. Wells would have been much impressed by this particular interpretation, but I nonetheless suspect he would have enjoyed the Honest Trailers summary of it.

3. Claire Tomalin, *The Young H. G. Wells: Changing the World* (Penguin, 2021), 6.

4. Tomalin, *Young H. G. Wells*, 25.

5. H. G. Wells, "The Man of the Year Million," *Tuapeka Times*, February 28, 1894, 5.

6. Wells, "Man of the Year Million," 5.

7. Tomalin, *Young H. G. Wells*, 49.

8. H. G. Wells, "The Rediscovery of the Unique," *Fortnightly Review* 50, no. 295 (July 1891): 106–111.

9. H. G. Wells, *The Time Machine* (London, 1895), 102–104.

10. Wells, *Time Machine*, 134–135, 139.

11. A. S. Eddington, *Internal Constitution of the Stars* (Cambridge University Press, 1926).

12. Tomalin, *Young H. G. Wells*, 76.

13. Wells, *Time Machine*, dedication.

14. Wells, *Time Machine*, 34.

15. H. G. Wells, "Fighters from Mars," *Boston Post*, January 10, 1898.

16. Wells, "Fighters from Mars," January 10, 1898.

17. Alexander MacDonald., *The Long Space Age: The Economic Origins of Space Exploration from Colonial America to the Cold War* (Yale University Press, 2017), 35–38.

18. Wells, *War of the Worlds*, 3

19. Wells, "Fighters from Mars," January 10, 1898

20. Wells, "Fighters from Mars," January 10, 1898.

21. H. G. Wells, "Fighters from Mars," *Boston Post*, January 11, 1898.

22. Wells, "Fighters from Mars," *Boston Post*, January 10, 1898.

23. Wells, "Fighters from Mars," January 10, 1898.

24. Wells, "Fighters from Mars," January 10, 1898.

25. Wells, "Fighters from Mars," January 10, 1898.

26. H. G. Wells, "Fighters from Mars," *Boston Post*, January 12, 1898.

27. Wells, "Fighters from Mars," January 12, 1898.

28. H. G. Wells, "Fighters from Mars," *Boston Post*, January 14, 1898.

29. Wells, *War of the Worlds*, 220.

30. H. G. Wells, "Fighters from Mars," *Boston Post*, February 2, 1898.

31. Wells, "Fighters from Mars," February 2, 1898.

32. Wells, "Fighters from Mars," February 2, 1898.

33. Wells, "Fighters from Mars," February 2, 1898.

34. Wells, "Fighters from Mars," February 2, 1898.

35. Wells, "Fighters from Mars," February 2, 1898.

36. Wells, "Fighters from Mars," February 2, 1898.

37. Wells, *War of the Worlds*, 301.

38. Wells, *War of the Worlds*, 301.

39. Wells, *War of the Worlds*, 301.

40. Garret P. Serviss, *Edison's Conquest of Mars* (Carcosa House, 1947), xi.

41. Clyde Fisher, "Garrett P. Serviss: One Who Loved Stars," *Popular Astronomy* 37, no. 7 (August–September 1929): 367.

42. Artemis Willis, "Performing Pictures: The Magic Lantern C. 1900" (PhD diss., University of Chicago, 2020).

43. Frank H. Winter, "The 'Trip to the Moon' and Other Early Space Flight Simulation Shows, Ca. 1901–1905," in *History of Rocketry and Astronautics*, vol. 23, ed. Donald C. Elder and Christophe Rothmund (Univelt, 2001), 135.

44. Willis, "Performing Pictures," 156.

45. Willis, "Performing Pictures," 161.

46. *Rocky Mountain News*, August 23, 1893, 4.

47. Serviss, *Edison's Conquest of Mars*, 3.

48. Serviss, *Edison's Conquest of Mars*, 5.

49. Serviss, *Edison's Conquest of Mars*, 7.

50. Serviss, *Edison's Conquest of Mars*, 8.

51. Serviss, *Edison's Conquest of Mars*, 9.

52. Serviss, *Edison's Conquest of Mars*, 13.

53. Serviss, *Edison's Conquest of Mars*, 16.

54. Serviss, *Edison's Conquest of Mars*.

55. H. G. Wells, *The World Set Free: A Story of Mankind* (E. P. Dutton & Company, 1914).

56. Serviss, *Edison's Conquest of Mars*, 24.

57. The total world GDP for 2024 was estimated by the World Bank at $111.25 trillion. See https://data.worldbank.org/indicator/NY.GDP.MKTP.CD.

58. Serviss, *Edison's Conquest of Mars*, 27; emphasis in the original. With these numbers, it is possible to calculate a per-spaceship cost for the fleet. The low end of the 2024 PWC ratio and GDP ratio equivalent numbers suggests a cost per ship, each with a complement of thirty disintegrator rays, of $28.5 billion and $157 billion per ship, comparable to the oft-cited cost of the International Space Station—that is, $100 billion. This makes this estimate arguably the first cost estimate in this category to within the right order of magnitude of eventual equivalent costs.

59. Serviss, *Edison's Conquest of Mars*, 34.

60. Serviss, *Edison's Conquest of Mars*, 35.

61. Serviss, *Edison's Conquest of Mars*, 34.

62. Serviss, *Edison's Conquest of Mars*, 36.

63. Serviss, *Edison's Conquest of Mars*, 36.

64. Serviss, *Edison's Conquest of Mars*, 37.

65. Serviss, *Edison's Conquest of Mars*, 43.

66. Serviss, *Edison's Conquest of Mars*, 44.

67. Serviss, *Edison's Conquest of Mars*, 45.

68. Serviss, *Edison's Conquest of Mars*, 49.

69. Serviss, *Edison's Conquest of Mars*, 51.

70. Serviss, *Edison's Conquest of Mars*, 54.

71. Serviss, *Edison's Conquest of Mars*, 66.

72. Serviss, *Edison's Conquest of Mars*, 66.

73. Serviss, *Edison's Conquest of Mars*, 76.

74. Serviss, *Edison's Conquest of Mars*, 98.

75. Serviss, *Edison's Conquest of Mars*, 106.

76. Serviss, *Edison's Conquest of Mars*, 111.

77. Serviss, *Edison's Conquest of Mars*, 119.

78. Serviss, *Edison's Conquest of Mars*, 121.

79. Serviss, *Edison's Conquest of Mars*, 136.

80. Serviss, *Edison's Conquest of Mars*, 137. Impressively, though it is believed today that Deimos is in fact a captured asteroid, this was not widely believed by astronomers at the time.

81. Serviss, *Edison's Conquest of Mars*, 166.

82. Serviss, *Edison's Conquest of Mars*, 85.

83. Serviss, *Edison's Conquest of Mars*, 185.

84. Robert Goddard, *The Papers of Robert H. Goddard*, vol. 1 (McGraw-Hill, 1970), 7.

85. Robert Goddard, "Diary, Oct 13, 1903," Robert H. Goddard Digital Collection, Clark University.

86. Garret P. Serviss, "Garret P. Serviss on Rockets to Mars," May 12, 1924, Robert H. Goddard Digital Collection, Clark University.

87. Robert Goddard, *The Papers of Robert H. Goddard*, vol. 2 (McGraw-Hill, 1970), 821–823.

88. Goddard, *Papers of Robert Goddard*, vol. 2, 825.

89. Garret P. Serviss, *The Moon Metal* (Harper & Brothers, 1900).

90. Garret P. Serviss, *A Columbus of Space* (D. Appleton, 1911).

91. Serviss, *Columbus of Space*, v.

92. H. G. Wells, *Tales of Space and Time* (Harper & Brothers, 1900).

93. H. G. Wells, *First Men in the Moon* (George Newnes, Limited, 1901), 68–69.

94. H. G. Wells, *In the Days of the Comet* (Century Co., 1906).

95. Wells, *World Set Free*, 303–304.

96. H. G. Wells, *The Outline of History* (Cassell and Company, 1920), 608.

97. *The Shape of Things to Come*, directed by William Cameron Menzies (London Film, 1936), 1:30:25–1:32:20.

CONCLUSION

1. Gerry Canavan, "'There's Nothing New / Under The Sun / But There Are New Suns': Recovering Octavia E. Butler's Lost Parables," *Los Angeles Review of Books*, June 9, 2014.

2. Wernher von Braun, *Project MARS: A Technical Tale* (Apogee Books, 2006).

3. Wendell W. Mendell, "Space Activism as an Epiphanic Belief System," in *Societal Impact of Spaceflight*, ed. Stephen J. Dick and Roger D. Launius (NASA, 2007), 573–583.

4. Kelly Weinersmith and Zach Weinersmith, *A City on Mars: Can We Settle Space, Should We Settle Space, and Have We Really Thought This Through?* (Penguin Press, 2023).

5. Rafe McGregor, "Octavia Butler's Parable of the Trickster," *Theaker's Quarterly Fiction*, April 14, 2025, https://theakersquarterly.blogspot.com/2025/04/octavia-butlers-parable-of-trickster.html.

6. Octavia E. Butler, *Parable of the Sower* (Grand Central Publishing, 2023), 85; emphasis in the original.

7. Butler, *Parable of the Sower*, 84–85.

8. Octavia E. Butler, *Parable of the Talents* (Seven Stories Press, 1998), 303; emphasis in the original.

9. Mike Wall, "U.S. Must Beat China Back to the Moon, Congress Tells NASA," Space.com, January 17, 2024, https://www.space.com/us-win-moon-race-china-congress-artemis-hearing.

10. Mary-Jane Rubenstein, *Astrotopia: The Dangerous Religion of the Corporate Space Race* (University of Chicago Press, 2022).

11. Rand Simberg, "Return of the Space Visionaries," *New Atlantis* 56 (Summer/Fall 2018): 48–68.

12. There is an argument to be made that it might be better to name this "space tribe" after Michel Ardan, the unashamedly space expansionist protagonist of Verne's *From the Earth to the Moon*. Unfortunately, to pronounce *Ardanists* in the anglicized way feels too far from the appropriate French pronunciation, and calling the space tribe *Les Ardanistes* in English also feels a step too far.

13. Robert J. Shiller, *Narrative Economics: How Stories Go Viral and Drive Major Economic Events* (Princeton University Press, 2020).

14. Konstantin Tsiolkovsky, "Is This Mere Fantasy?," in *The Science Fiction of Konstantin Tsiolkovsky*, ed. Adam Starchild (University Press of the Pacific, 1979), 453–455.

15. Michael J. Neufeld, *Von Braun: Dreamer of Space, Engineer of War* (Vintage, 2008), 2; and James Harford, *Korolev: How One Man Masterminded the Soviet Drive to Beat America to the Moon* (John Wiley & Sons, 1997), 8, 12–14.

16. William Sims Bainbridge, *The Spaceflight Revolution: A Sociological Study* (John Wiley & Sons, 1976), 38.

17. Bainbridge, *Spaceflight Revolution*, 38.

18. Joel P. Flynn and Karthik Sastry, "The Macroeconomics of Narratives," NBER Working Paper 32602 (National Bureau of Economic Research, June 2024), https://doi.org/10.3386/w32602.

INDEX

Page numbers in italics refer to figures and tables.

Publisher contact:
The MIT Press
Massachusetts Institute of Technology
77 Massachusetts Avenue, Cambridge, MA 02139
mitpress.mit.edu

EU Authorised Representative:
Easy Access System Europe, Mustamäe tee 50,
10621 Tallinn, Estonia
gpsr.requests@easproject.com

Printed by Integrated Books International,
United States of America